CONTENTS

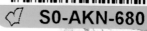

S0-AKN-680

Through millennia waves and currents sculpture the coasts of the world. Here they are working in volcanic rock. (Seal Rocks St. Pk., Oreg.)

LOOKING AT LANDFORMS

Mountains, valleys, and plains seem to change little, if at all, when left to nature. We call them "everlasting." But they do change—continuously. Features of Earth's crust are but temporary forms in a long sequence of change that began when the planet originated billions of years ago, and is continuing today.

The processes that shaped the crust in the past are shaping it now. Understanding them, we can imagine in a general way how the land looked in the distant past and how it may look in the distant future. We begin to see this Earth in a new dimension.

Landforms are limitless in variety. Some have been shaped primarily by streams of water, others by glacial ice, still others by waves and currents, and many more by slow convulsions of Earth's crust or by volcanic eruptions. There are landscapes typical of deserts, and others characteristic of humid regions. The arctic makes its special marks on rock scenery, and so do the tropics. Because geological conditions from locality to locality are never quite the same, every landscape is unique.

For an interested observer, landforms always raise questions. Why are the hills and valleys where they are? Why does the river take that turn? Why is this region riddled with caves, while others are not? Why does the valley have a steep wall on one side, a gently sloping wall on the other? What makes this terrain a badland, that one a land of rolling hills? It is only natural that man, spending his lifetime among landforms, should ask such questions. This book will provide some of the answers.

Beds of volcanic ash are eroded by rainwash into weird forms. Hard lava fragments litter foreground. (Big Bend Nat. Pk., Texas)

Weather, running water, and glaciers carve mountain masses into horn peaks, deep valley-troughs, and basins for sparkling lakes. (Lake Solitude, Teton Range, Wyo.)

HOW TO USE THIS GUIDE

The purpose of this book is to help you identify and understand common kinds of natural rock scenery. Identification and understanding depend on an acquaintance with the nature of Earth's crust and the geologic processes that shape it. These fundamentals are described in the book's earlier sections; later sections show how geologic processes make particular kinds of scenery under special conditions. Photographs have been chosen to illustrate as nearly as possible what is "typical"—i.e. the forms most likely to be seen by an alert, observing traveler. Basic processes and concepts are explained by means of diagrams where necessary. Finally there are suggestions on using topographic maps and on finding and using sources of information beyond the scope of this book.

READ THIS GUIDE, particularly pp. 10-59, in sequence. In this way basic concepts and terms will be learned in logical order. Once these are understood, skipping back and forth in the book can be more interesting and informative. Frequent browsing will sharpen your alertness to landform characteristics.

USE THE INDEX OFTEN. It will direct you not only to key definitions and explanations, but also to related text and illustrations on other pages.

CARRY THIS GUIDE when you travel—especially when visiting places where rock scenery is spectacular and varied. Mountains are likely to be most striking, but flat country is worth observing too, whether it be plain, plateau, or seashore. Look for streambeds, roadcuts, steep-sided hills, cliffs, excavations—any sites where bedrock is exposed. Note shapes of hills and valleys and the patterns they form. (Drivers: caution!)

USE SOURCES OF INFORMATION about local landforms wherever you are. Check your library or write the appropriate agency of your state (p.155) for books, maps, leaflets. Before leaving on an extended trip get information about landforms to watch for en route.

OBSERVE INTENSIVELY Much can be seen from a car, bus, train, boat, or airplane. But intensive looking usually involves some walking—perhaps hiking or even mountain climbing. When visiting national or state parks join nature walks led by rangers, who will point out features of geologic interest. There may be in your home area a hiking group that makes trips to scenic places. People who can walk a few miles will see interesting features inaccessible to those who always ride.

Hikers inspect a glacier under guidance of a National Park ranger. (Glacier Nat. Pk., Mont.)

Geology students on field trip investigate structure and composition of a rock outcrop.

STUDY LANDFORMS closely. They are less easily identified than plants and animals. Plains vary more than ducks, volcanic necks more than daisies. Identification may hinge not only on general appearance but on rock structure and composition, and on which agents of change have been dominant. The more geology we know, the better we can understand rock scenery.

USE MAPS as aids to intensive observation. Topographic maps of most parts of the United States, on a scale of one or two inches to the mile, with 20-ft., contour intervals, are available from the U.S. Geological Survey (p. 155). Such maps show clearly the "lay" of the land—relative sizes and positions of hills and valleys, elevations, courses of streams, locations of swamps and ponds. To a geologist "topo" maps offer strong clues to subsurface rock structures that have influenced the shaping of the land. For use of maps see p. 154.

USE YOUR CAMERA to increase your alertness to landforms and the opportunities for studying them. Getting the picture that will best show the essential nature of the feature, with interesting composition and lighting, is a challenge. Often it involves observation from several angles, and it may result in the discovery of important details that otherwise would be missed. A good picture in your file not only raises pleasant memories but is a scientific record of what was seen.

SEEK FURTHER INFORMATION about landforms in recommended textbooks, technical articles, and popular presentations. (Page 155 lists some of the best from the viewpoint of readability, reliability, and illustrations.) You may also consider taking a course in geology at a nearby college or adult school.

Alcove cut into valley wall by torrential stream during flood periods offers rare opportunity to a photographer. (Hocking St. Pk., Ohio)

EARTH'S CRUST

All landforms are shaped in and on Earth's crust. This half-rigid "skin" of rock is thickest (up to 35 or 40 mi.) beneath the continents and thinnest (as little as 3 or 4 mi.) beneath the ocean. It covers the mass of plastic, rocklike material that forms the mantle. The mantle encompasses the outer core, which consists probably of iron and nickel, mostly in liquid form. Within this is the inner core, a solid ball of nickel and iron. The inner structure of the planet has been inferred mostly from observations of seismic (earthquake) waves.

Considering that Earth's diameter is about 7,926 mi., the surface of the crust is relatively smooth. Only about 12 mi. separates the highest elevation (Mt. Everest, 29,208 ft.) from the lowest (Mariana Trench, southwest Pacific, 35,800 ft.). Oceans cover the lower levels —over 75% of the planet's surface. Continents are formed mostly by rock masses that are a little lighter and thus stand slightly higher than the rock masses

Earth is made of concentric spherical masses. Their existence is inferred from analysis of earthquake (seismic) waves which penetrate the interior. Thus, in the diagram, arrows from the epicenter E represent waves which pass only through rigid solids. These are received at seismic stations S_1, S_2, S_3 on the crust. Beyond S_3 these waves are not received; hence a fluid obstacle is inferred.

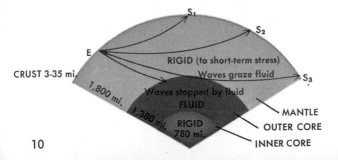

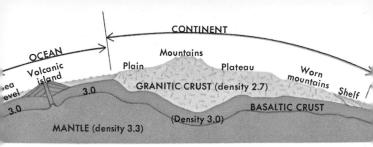

Elevations of major parts of crust depend on their density. Continental blocks, lighter than oceanic blocks, stand higher.

whose tops form the ocean bottoms. The topography of these bottoms resembles that of continents: mostly plains and gently rolling surfaces, here and there interrupted by volcanic hills or mountain systems.

The crust is shaped by three groups of processes: *solid movements,* involving distortion, breakage, and dislocation of rock; *igneous activity,* or generation and movement of molten materials; and *gradation,* or leveling of land by disintegration of crustal rock and spreading of the waste (see pp. 14-16).

Balance in crust is maintained by "isostatic" movements. Thus as mountain blocks lose mass by erosion they tend to rise, and as ocean blocks are weighted with erosion debris they sink. Such motions deform and fracture the crust (see p. 14).

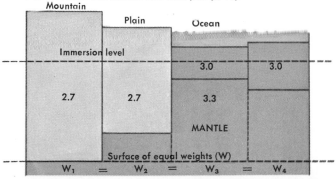

PHYSIOGRAPHIC REGIONS OF THE UNITED STATES

1. Atlantic Shelf: sloping submarine plain of sedimentation from shoreline to 600-ft. depth.

2. Coastal Plain: low, hilly to nearly flat belted and terraced plain on soft sediments.

3. Blue Ridge-Piedmont: complex mountain structures eroded low on east, remaining higher on west (Blue Ridge Mts); rounded summits 3,000-6,000 ft.

4. New England: glaciated equivalent of Blue Ridge-Piedmont; Blue Ridge replaced by sharper peaks of White and Green Mts.

5. Valley and Ridge: long, parallel mountain ridges eroded from regular folds.

6. Appalachian Plateau: generally steep-sided, deeply dissected; sandstones predominating; 2,000-4,000 ft. on east, lower toward west.

7. Adirondacks: complex mountains on ancient crystalline rocks: some summits over 5,000 ft.; wide areas of moderate altitudes and relief.

8. Interior Low Plateaus: open, well-dissected domes and basins; low escarpments separating more level stretches.

9. Ozark Plateaus: high, hilly; on stratified rocks.

10. Ouachita Mts: parallel ridges and valleys eroded from folded strata.

11. Central Lowlands: low, rolling plains; mostly veneer of glacial deposits, including lake beds and lake-dotted moraines; unglaciated southwestern part shows flat sediments little deformed or dissected.

12. Superior Upland: like Adirondacks, but mountains lower; broad alluvial plains reaching 6,000 ft. near Rockies.

13. Great Plains: erosion in north exposes stratified and igneous rocks related to Rocky Mt buildup.

14. Central and Southern Rockies: broad anticlines eroded down to crystalline basement rock, which preserves ranges and peaks to 14,000 ft.; broad synclinal basins separating anticlines; extrusive volcanic rocks forming dissected plateaus in northwest part.

15. Northern Rockies: complex mountains with narrow intermontane basins.

16. Columbia Plateau: high rolling plateaus on extensive volcanics; deeply canyoned.

17. Colorado Plateau: high plateau on sedimentary rocks, deeply canyoned.

18. Basin and Range: mostly inland fault block ranges separated
by wide desert plains; many ancient lake plains and alluvial fans.

19. Cascade-Sierra: Sierras, in
south, eroded from huge fault block
in granitic rocks; Cascades, in north,
are warped volcanics surmounted
by high volcanic cones.

20. Pacific Border: young fold-fault
mountains flanked on east by extensive river plains in California
section, on west by structurally active shelf area.

☐ Cordillera
☐ Central Plains
☐ Interior Highlands
■ Shield Areas
☐ Appalachians
☐ Atlantic Plain

13

LAND-SHAPING PROCESSES

SOLID MOVEMENTS   In the continents large portions of the crust move both horizontally and vertically. Heat in the mantle is believed to set up a slow convection current or a "blister" which raises the upper part of the mantle and then drags it sidewise, carrying overlying crust with it. Vertical movements in the crust occur as erosion debris on the surface shifts from block to block and as masses of water are shifted by growth and melting of glaciers.

Distortion of crust is shown by tilt of strata that were originally horizontal. (San Rafael Swell, Utah)

IGNEOUS ACTIVITY   Heat is produced in mantle and crust by radioactivity and by friction between moving crustal blocks. This heat melts some crustal rock and helps to keep the mantle partly fluid or plastic. The weight of the crust causes the molten material to squeeze up through crustal fractures. Some cools and solidifies within the fractures; some bursts through the crust as volcanic eruptions. This hot, fluid material is called *magma* while it is below the surface, *lava* during and after eruptions. When solidified in or on the crust, it is *igneous rock*.

Earth's crust, oceans, and atmosphere were formed probably four to five billion years ago as molten material welled up and cooled. The first landscapes were eroded away long ago, but continued igneous activity has been making new ones ever since.

14

Recently active volcanic area suggests how most of Earth's surface may have appeared when crust was forming. In foreground, lava and spatter cones; cinder cones beyond. (Craters of Moon, Idaho)

GRADATION   Some land is lowered by destruction of rock and removal of the resulting rock waste. This entire process is called *degradation*. Where the rock waste is deposited, the elevation of the land surface is raised—a process called *aggradation*. The two complementary processes, considered together, are *gradation*. In its many aspects gradation includes weathering, mass wasting, and erosion and deposition.

Weathering is the disintegration of rock as a result of exposure to conditions on Earth's surface. Atmospheric gases and moisture attack rock chemically, converting it into compounds that are usually more stable but mechanically weaker. Acids from decaying organic matter attack rock. Water freezing in crevices expands, forcing rock masses apart; growing tree roots do the same. Contraction and expansion due to temperature changes may cause rock to break. As rock formed at depth becomes exposed by erosion and relieved of confining pressure, it may expand and crack.

15

Rock broken by weathering descends to form talus cone at foot of steep ravine. (Teton Range, Wyo.)

Mass wasting is the descent of loose material to lower levels by the action of gravity. Weathering loosens rock on cliffs, and fragments fall. Masses of fragments on steep slopes may be disturbed by ice-prying or heavy rain; then a landslide or earthflow starts. On gentle slopes rock material descends gradually.

Erosion is the wearing away of rock by active agents which also transport and eventually deposit the waste. Of these agents the most erosive are streams of water, which cut into rock by abrading it with sediments and decomposing it, and thus make valleys. Water underground dissolves rock, making cavities. Wind "sandblasts" rock and transports loose material, forming dunes. Waves and currents erode shores and pile up sediments to make bars and beaches. Glacier ice, creeping over arctic lands or down mountain valleys, disintegrates bedrock by grinding and plucking, and thus reshapes highlands and lowlands.

GEOLOGIC TIME Once men believed hills, valleys, and other landforms were created by cataclysms long ago and have changed little since. But the pioneer Scottish geologist James Hutton (1726-97) proposed that the agents which shaped landscapes in the past were much the same as the agents at work today. They

Grand Canyon of the Colorado is a demonstration of geologic time (see Geologic Time Table, p. 18.) Rock layers increase in age from top (Permian) to bottom (Precambrian) of canyon, spanning a billion years. Upper part of picture shows famous Algonkian Wedge (pointing left), an "unconformity" formed by horizontal Cambrian strata overlying tilted and beveled Precambrian strata which are much older.

are slow, but geologic time—the billions of years during which geologic agents work—is long.

Hutton's principle, called *uniformitarianism,* became a foundation of geology. By studying geologic changes now going on, scientists have been able to reconstruct much of Earth's past (see timetable, p. 18).

The pace of change varies. Volcanic action can build up a small mountain in a year or two. The Mississippi made its delta in a million years, and the Colorado River took about two million to cut the Grand Canyon. Lofty mountains like the Rockies may take ten million years to rise and two or three times as long to erode down to low hills. Such figures suggest the vastness of geologic time.

A few erosional forms on present landscapes began taking shape as far back as 50 million years ago. Most have developed during the past few million years. Many spectacular features have emerged with the waning of the Pleistocene glaciers (pp. 112-120). In many landscapes the rock is much older than the present phase of sculpturing.

THE GEOLOGIC TIME CHART

Era	Period	Epoch	Events with special reference to landforms	Millions of Years Before the Present
CENOZOIC — Time of recent life and modern landscapes	Quaternary — Time of recent glaciers and man	Recent	Glaciers partly melt; sea level rises; modern deserts appear. Man modifies the earth.	
		Pleisto-cene	Extensive glaciers. Widespread erosion controlled by fluctuation of sea level with the waxing and waning of glaciers. Rise and deformation of Himalayas continue.	2
		Pliocene	Uplift in older and younger mountain areas; renewed valley cutting; deformation along Pacific Coast begins.	12
	Tertiary — Intense mountain-building and volcanism	Miocene	Culmination of Alpine-Himalaya movements; extensive leveling by erosion in Appalachians and Rockies.	25
		Oligocene	Crustal movements begin forming Alps and Himalayas. Deposition continues in Rocky Mountain basins. Faulting in Basin and Range Province.	40
		Eocene	Sediments accumulate in Rocky Mountain basins. Ending of worldwide warm, uniform climate.	60
		Paleocene	Structures of Rocky Mountains completed. General framework of modern landscapes begins to emerge.	70
MESOZOIC — Time of dinosaurs and vanished scenery		Cretaceous	Forming of Rocky Mountain structures begins.	135
		Jurassic	Earliest evidence of continental drift; volcanism in western U.S.	170
		Triassic	Rift valleys and volcanism in eastern U.S. and Canada.	225
PALEOZOIC — Early life in sea and on land		Permian	Broad icecaps on southern continents.	270
		Carbonif-erous	Appalachians and pre-Alpine structures forming.	350
		Devonian	Vegetation begins to cover land.	400
		Silurian	Earliest known coral reefs.	460
		Ordovician	Abundant marine life—fossils aid in determining age of rocks.	500
		Cambrian		
PRECAMBRIAN — Beginnings of life		Late	Earliest glaciation.	600
		Early	Earliest record of present geologic processes—deformation, volcanism, gradation.	3500

ROCK: MATERIAL OF LANDFORMS

BEDROCK AND MANTLE   Earth's crust consists of a massive, hard rock layer covered here and there by loose material. The hard layer, called *bedrock,* is exposed in cliffs and roadcuts, on rocky shores and mountain summits. The loose material, called *mantle rock,* or "mantle," consists mostly of disintegrated rock. (It is not to be confused with *the* mantle—the zone beneath the crust.) When mixed with plant and animal wastes mantle becomes *soil.* This is "residual" if it lies where it formed, "transported" if it has been moved, e.g. by a glacier. Mantle tends to move downslope by mass wasting, and thus accumulates in depressions and where slopes are gentle.

ROCK COMPOSITION   Rocks are mixtures of mineral grains naturally consolidated. Minerals are chemical compounds with definite crystalline forms. Of the 2,000 varieties recognized, only a handful form the bulk of all rocks. Each rock consists mainly of two or three kinds of the minerals described on p. 20. More detailed information about minerals can be found in the guidebooks listed on p. 155.

Quarry shows mantle of loose material overlying slate. Slate is the "country rock," or bedrock generally underlying the area. (Pennsylvania)

COMMON ROCK-FORMING MINERALS

Name	Composition	Identification			
		Color	Hardness	Luster	Mode of Fracture
Carbonates: calcite	Calcium carbonate	Colorless if pure; varied mixtures	Equal to copper coin	Bright-glassy	Three acute-angle flat breaks
dolomite	Calcium, magnesium carbonate				
Clay minerals	Complex hydrated aluminosilicates	White if pure; easily stained	Less than finger-nail	Dull	Earthy (greasy feel)
Feldspars: orthoclase	Aluminosilicates of: potassium	Buff, pink	Like steel	Bright-glassy	Two right-angle flat breaks
plagioclase	sodium and calcium	white, gray			
Ferro-magnesians	Complex alumin-osilicates of iron and magnesium	Green to black	Less than steel	Glassy	Two acute-angle flat breaks
Gypsum	Calcium sulphate	Colorless if pure; often white, pink	Less than finger-nail	Bright-glassy	Three acute-angle flat breaks
Iron oxides: limonite	Hydrated ferrous oxide	Yellow-brown powder	Varies	Dull or metallic	Irregular
hematite	Ferric oxide	Red-brown powder			
Micas: muscovite	Hydrated aluminosilicates of: potassium	Colorless	Less than copper coin	Pearly	Peels
biotite	iron or magnesium	Brown, black			
Olivines	Silicates of iron and magnesium	Olive greens	Greater than steel	Glassy	Curved breaks
Quartz: crystalline	Silicon dioxide (silica)	Colorless if pure; many colors	Greater than steel	Glassy or greasy	Unequal curved breaks
chert, flint				Relatively dull	Smooth curves
Serpentines: antigorite chrysotile	Hydrous magnesium silicates	Mottled green	Less than copper coin	Greasy or silky	Irregular or fibrous

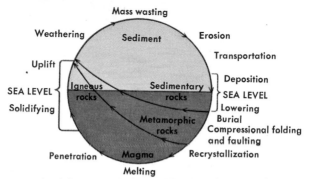

Igneous rocks (left) are made by solidification of magma derived from deep-seated melting. Uplift above sea level exposes rock to weathering, mass wasting, and erosion. Resulting sediments are transported and deposited, and may become sedimentary rock. This when lowered may become buried, converted into metamorphic rock, and melted to form newer magma. Sedimentary and metamorphic rocks may, like igneous rock, become subject to uplift and erosion.

ROCK VARIETIES vary from region to region. The kinds of minerals that form them, and the ways in which the minerals are arranged, strongly influence the nature of landforms shaped by erosion (pp. 28-29). The three general categories of rock are *igneous*, *sedimentary*, and *metamorphic*. The types most commonly encountered as landforms are described on pp. 22-27.

Modes and places of origin of rock types: 1 sea bottom sediments; 2 basic crust; 3 metamorphics; 4 shelf sediments; 5 sill; 6 shore sediments; 7 dike; 8 young batholith; 9 old lava flows; 10 metamorphics; 11 old batholith, 12 intermontane sediments; 13 lava cone and flows; 14 folded sediments of all types.

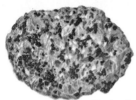
Pink granite (New England)

Diabase, a gabbro (New York)

Igneous rocks form by cooling and solidification of magma in the crust, and of lava on the crust. Rock formed within the crust, called *intrusive,* is generally heavy and hard. Rock formed on the crust, known as *extrusive,* may range from heavy and hard to light and crumbly or powdery. Igneous rocks are found where the crust has been fractured (see pp. 60-74).

GRANITE (intrusive): visible interlocking crystalline grains, mostly orthoclase feldspar and quartz; often mica or ferromagnesian minerals. (Rock is *syenite* if no quartz.) Overall hardness of steel; light-colored. Relatively resistant; batholiths (p. 72) often form mountain cores.

GABBRO (intrusive): visible interlocking crystalline grains, mostly plagioclase feldspar and ferromagnesian minerals; perhaps olivine; no quartz. Overall hardness of steel; dark. Ferromagnesians weather relatively fast in humid areas. In dikes, sills, batholiths (pp. 72-74).

Widely spaced joints (fractures) divide granite into large blocks, which become separated by weathering. (Hudson Highlands, N.Y.)

Rhyolite eroded into pillars (Lewis River Canyon, Wyo.)

Basalt cliff with varied jointing (Columbia Gorge, Wash.)

FELSITE (Incl. rhyolite) (extrusive): microscopic interlocking crystalline grains, mostly orthoclase feldspar, quartz; often mica, ferromagnesians. Overall hardness of steel; light-colored. Occurs mostly as lava flows and pyroclastics (pp. 66, 68). Common variety is rhyolite.

BASALT (extrusive): usually microscopic interlocking crystalline grains; much plagioclase feldspar; more ferromagnesians than in andesite. Over-all hardness of steel. Dark green to blue-gray. Occurs mostly as lava flows and pyroclastics (pp. 66, 67). Joints, due to contraction

after solidifying, may form prismatic columns (p. 67). Weathers brownish. May have small gas-produced cavities (amygdules) that filled later with other kinds of minerals.

ANDESITE (extrusive): mixed microscopic and visible interlocking grains; mostly plagioclase feldspars, ferromagnesians. Overall hardness of steel; medium shades of red and green; gray. Occurs mostly as lava flows and pyroclastics (pp. 66, 95-96).

Note: For other forms of igneous rock see pp. 66-68.

Rhyolite (Wyoming)

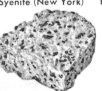

Syenite (New York)

Basalt (New Jersey)

23

Limestone of massive, resistant type (Vercors, France)

Shell limestone—a variety rich in fossils

Sedimentary rocks make up most top layers of the crust, because this is where weathering, erosion, and deposition occur. These rocks may be found almost anywhere. They are made of (1) loose mineral particles deposited on land or in water, then compacted and cemented, e.g. sandstone from sand deposits and shale from mud; (2) crystallized precipitates from sea water, e.g. some limestone. Sedimentary rocks are usually in layers, or strata, also called *beds,* separated by *bedding planes.*

LIMESTONE: mostly calcite or dolomite. Crystals or fragments of organic (shells), chemical (precipitates), and mechanical origin. White to blue or gray. Hardness equal to copper coin (harder if cemented with silica). Subject to solution, forming sinks, caves, valleys, especially in humid climates (pp. 130-140). May form highlands in arid or recently elevated regions.

Shale of common type—thin-bedded, crumbly (New York)

Red shale—a variety common in many states

Conglomerate and sandstone interbedded (New Jersey)

Massive sandstone eroded into various forms (New Mexico)

SHALE: mostly clay minerals and quartz, colored by carbon, metallic oxides, and ferromagnesians. From muds and silts. Fine-grained, thin-bedded; flaky or slabby; almost any color, especially red, brown, or black; soft unless infused with silica or baked by magma or lava. Generally relatively weak, forming valleys not ridges.

SANDSTONE: mineral grains of quartz, feldspars, ferromagnesians, gypsum, etc.; and rock grains of granite, basalt, limestone, etc. Grains of sand size, deposited by streams near highlands or waves and currents near a shore, or by winds.

Grains compacted when buried, and cemented by silica, calcite, iron oxide, etc. Red, brown, green, yellow. Rock often resistant to weathering and erosion but succumbs to undermining and mass wasting, which cause retreat of cliffs.

CONGLOMERATE: sand and rounded pebbles naturally cemented. Deposited by streams at foot of highlands or waves and currents on beaches. Erodes like sandstone.

BRECCIA: jumbled angular rock fragments naturally cemented. Along some faults, at base of cliffs, and in submarine slides.

Ripple marks made in sand that later turned to stone (Front Range, Colo.)

Crossbedding in windblown sand that turned to stone (Walnut Canyon, Ariz.)

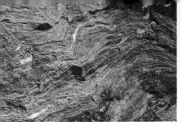

Schist intensely folded during metamorphism—a common phenomenon (Green Mts, Vt.)

Cliff of quartzite with typical blocky aspect due to rectangular jointing (Green Mts, Vt.)

Metamorphic rocks are igneous or sedimentary rocks changed (metamorphosed) by internal heat, pressure, and penetration by fluids, without melting. They form deep in the crust, especially in zones of mountain building (p. 87), and are exposed later by erosion. Most are hard and, except for marble, chemically resistant.

QUARTZITE: derived from a quartz sandstone. Breaks through quartz grains (not around them) because they are cemented with quartz from ground water. Shows smooth, often "sugary," homogeneous surface. Dense and tough; usually most resistant rock in locality.

SCHIST: derived from shale or igneous rock. Visible interlocking crystalline grains mostly of mica, hornblende, quartz; arranged in sheets, along which rock breaks. Sheets may not be parallel to bedding; often contorted. Hardness increases with quartz content. Rock may be flaky; generally resistant.

Quartzite—a specimen with the characteristic sugary look; common in many states

Mica schist—one of the commoner types, less shiny and less flaky with mica than some

Gneiss with hornblende bands, dislocated during metamorphism (Hudson Highlands, N.Y.)

Slate flag with edges of bedding planes of original shale appearing as bands (Penna.)

SLATE. derived from shale. Microscopic interlocking crystalline grains, mostly mica, quartz, clay; in sheets, along which rock breaks. Sheets may not parallel bedding. Softer than steel. May grade laterally into shale (less lustrous). Resistant.

PHYLLITE: derived from shale. Like mica schist in composition and appearance, but often more lustrous, silvery. Flaky; sheets may not parallel bedding. May grade laterally into schist (more metamorphosed) or slate (less metamorphosed). Resistant.

GNEISS: derived from shale, sandy shale, or igneous rocks. Visible interlocking crystalline grains, mostly feldspar with some quartz and streaks or bands of mica and hornblende. May grade into schist. Often blocky or layered. Generally resistant, often forming highlands.

MARBLE: derived from limestone. Usually visible interlocking crystalline grains, mostly calcite and dolomite. White when pure; often streaked or mottled with other minerals. Hardness equal to copper coin. Typical valley-former in temperate humid climates. May form peaks in young mountains.

Marble with typical dense texture and layers of different hues (Green Mts, Vt.)

Marble cliff showing strata severely crumpled by mountain-building forces (Green Mts, Vt.)

Structural Influences on a Coastal Plain

RESISTANCE OF ROCKS to degradation varies, as indicated in descriptions of rock types on pp. 22-27. Some rocks (e.g. basalt) consist of minerals that are relatively unstable chemically in a given climate; others (e.g. granite) consist mainly of more stable minerals and thus are chemically more resistant. Relatively hard rocks (e.g. quartzite) resist mechanical attack (e.g. stream abrasion) more effectively than softer rocks (e.g. shale). Differences in resistance mean differences in rates of degradation; thus on many landscapes the higher elevations are on "strong" rocks, the lower ones on "weak" rocks.

STRUCTURES OF ROCKS vary widely. Minerals making up rock are arranged in different ways (e.g. in clusters or layers). Rock masses have various positions

Butte and mesa: from horizontal strata (New Mexico)

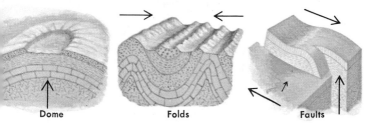

Dome   Folds   Faults

Topography on various structures: Arrows indicate earth movements.

with respect to each other (e.g. over or under, or tilted or not tilted). Rock layers may be folded or warped.

All rock is more or less fractured, or broken. Commonly, fractures result from earth movements or contraction of igneous rock as it cools. Fractures that involve no displacement of the adjacent rock masses are *joints*. Fractures that do involve displacement are *faults*. Joints and faults are important aspects of structure.

Structures guide the shaping of relief. Valleys tend to be cut in belts of weaker rock or along major fractures. Ridges and knobs develop on stronger rock masses. Highlands with flat tops tend to be produced by erosion of horizontal strata, and sharp-backed ridges ("hogbacks") by erosion of the edges of resistant strata that are steeply tilted. Erosion of a dome may make a ring of hills. The accompanying diagrams show some typical expressions of structure.

Strike and dip describe attitudes of sloping planes in rock structures. Strike is the compass direction of any horizontal line in the plane. Dip is the vertical angle between the plane and any horizontal line drawn perpendicular to the strike line Here the strike is about N45°W.

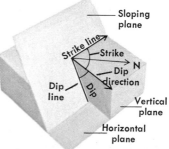

Water freezing in cracks breaks off slabs of rock. (New Jersey)

Tree roots growing in crevice can split boulder. (New York)

THE WORK OF WEATHERING

Rock at or near the surface of the continents breaks up and decomposes because of exposure. The processes involved are called *weathering;* some are mechanical, others chemical. Weathering produces some landforms directly, but is more effective in preparing rock for removal by mass wasting and erosion (p. 16). Weathering influences relief on every landscape.

FREEZING AND THAWING Water freezes to form crystalline ice, which occupies more space than the water. The expansive force involved in freezing approaches 2,000 lb. per sq. in. Expansion in rock crevices can break off fragments or push rock masses apart. This *ice-prying* is most destructive on arctic and mountain terrains, where thaw-freeze is frequent. It produces much of the rock litter there.

As wet soil freezes, ice crystals ("frost flowers") raise the soil and break it up. "Frost-heaving" of pavements is similar. In arctic lands freezing of former lake bottoms may raise soil-capped mounds, called *pingoes,* as high as 100 ft. Wide arctic areas become divided by ice partitions into *polygonal (patterned) ground.* Repeated thawing and freezing sort rock debris, often forming *stone rings* of large stones around concentrations of small ones.

TEMPERATURE CHANGES Sudden cooling of a rock surface may cause it to contract so rapidly over warmer rock beneath it that flakes or grains break off. This happens mostly in deserts, where intense daytime heat is followed by rapid cooling after sunset or chilling by sudden rain. Ordinary temperature changes are, however, less destructive than was once supposed.

ORGANIC ACTION Plant roots expand as they grow, breaking up soil and moving small rock masses apart. Animals such as prairie dogs and moles break up soil by burrowing. This weathering by organic action is greatest in warm, humid regions rich in plant and animal life.

Pingoes on a former lake bottom (N.W. Territories, Canada)

Patterned ground developed in a broad tundra area (Alaska)

Flakes and grains are broken off granite by chemical weathering and ice-prying. Thick slabs—about 3 in. or thicker—break off because of unloading (p. 34). (Hudson Highlands, N.Y.)

CHEMICAL ACTION Minerals in rock decompose if long exposed to atmospheric water and dissolved gases. Decomposition, most rapid in warm, humid regions, produces new minerals, usually mechanically weaker and of larger volume than the original minerals. The rock either crumbles or *exfoliates* (flakes off).

Sandstones and shales, consisting mainly of previously weathered rock waste, resist chemical weathering. Limestone, marble, and gypsum are dissolved by water containing carbon dioxide (pp. 130-140).

Water and carbon dioxide react with feldspars, the most abundant minerals in most igneous rocks, to make clay and silica. Water alone may react with ferromagnesians to produce limonite and hematite—the sources of most yellow and reddish colors in rock.

DIFFERENTIAL WEATHERING If the minerals in an angular block have about the same resistance to weathering, or differ in resistance but are evenly distributed, prolonged weathering will round off the block. The reason is that exposure is greater along the edges. Granite, usually homogeneous, thus often weathers to rounded boulders.

Rock containing minerals of varying resistance, seg-

Spheroidal weathering of granite below ground surface exposed by roadcut (Colorado)

Granite masses rounded by prolonged weathering (Devil's Marbles, Australia)

regated into layers or pockets, weathers unevenly as these become exposed. Edges of the more resistant layers stand out as ridges, and pockets of resistant mineral as humps or knobs. This aspect of weathering is called *differential*. Along with erosion, which also has this aspect, it accounts for much landscape relief.

Weathering is guided by joints (p. 29), which offer access to air, moisture, and other weathering agents. These penetrate scores or even hundreds of feet into bedrock. The closer the jointing, the faster the weathering, because more surface is exposed. Predominantly vertical, closely spaced joints favor weathering to pillars and needles; horizontal joints favor slabs or sheets; rectangular joints yield blocks. (See p. 26.)

Sculpturing of "needles" and "fins" guided by joints that are close and also mainly vertical (West Virginia)

"Owl Rock": showing effects of rather widely spaced vertical jointing and horizontal bedding (Arizona)

UNLOADING Rock formed at depth under high pressure expands as erosion removes, or "unloads," overlying rock. Expansion may cause fracturing parallel to the topographic surface, producing parallel "plates" or "sheets" from a few inches to a few feet thick. On level or gently sloping topography, a set of sheets in structureless rock like granite may look much like bedding planes. Where valleys and troughs are cut into granite masses, sheets tend to parallel valley walls. Thus, in older terrains, sheets are rounded. Rounding may persist until only a few knobs remain; these are called *exfoliation domes*. The Sierra Nevada in California, the Colorado Rockies, and the older Appalachians offer examples.

Granite mountain has been rounded by exfoliation due largely to unloading. Note lack of joints other than those associated with the exfoliation shells. (Dome Rock, Sequoia Nat. For., Calif.)

Fragments of quartzite broken off cliff by weathering accumulate to form the apronlike talus slope at its foot. (Vermont)

FEATURES OF MASS WASTING

Rock material loosened by weathering is continuously pulled by gravity. If support is removed, the material tilts, falls, slides, flows, or sinks. These movements from gravity alone are called *mass wasting*. Some are dramatic and destructive, e.g. huge landslides triggered by earthquakes. But mass wasting, like weathering, is mostly slow, and usually it does not alone produce large, spectacular landforms. Naturally, mass wasting is most effective in highlands, where slopes are steep.

ROCKFALLS are falls of individual rock fragments from a cliff. Fragments range from small flakes to blocks weighing hundreds of tons, which tend to break on the way down. An accumulation of fallen material at the foot of a cliff is a *talus*. Falls of mantle rock, often mixed with vegetation, are *debris falls*.

Rockslide crossed valley bottom and climbed up opposite slope. River was blocked to form lake. (Madison River Canyon, Mont.)

ROCKSLIDES are movements of huge masses of bedrock down a sloping structural surface, e.g. a tilted bedding or joint plane. The cause may be lubrication or weakening of the mass by heavy rains or undermining. Loosened material may accelerate to 90 m.p.h., descend into a valley, break up, and climb the opposite wall. Slides can wipe out towns; some block a stream to make a lake. Slides of mantle rock mixed with some vegetation are *debris slides*.

SLUMPS are slides or slips along curved surfaces within rock masses that lack joints or bedding planes. The slip produces backward rotation. Slumping usually occurs in weak hillside material, e.g. clay or sand, when saturation by water lowers its shearing strength. Streams can cause slumping by undermining their banks. Bedrock masses sometimes slump; huge rotated *slump blocks* result (see facing page).

Slump block (left) of gypsum in roadcut (New Mexico)

Avalanche chutes on precipitous slope (Front Range, Colo.)

EARTHFLOWS are movements at the toe of a slump block that is not in contact with a stream or other erosive agent. Movement results from chaotic internal sliding, which produces a hummocky surface

AVALANCHES are swift downflows through steep, narrow channels. Snow, ice, or rock debris, or any combination of these, may be involved. Typical alpine avalanches are mainly heavy snow accumulations which, after being generally weakened by overweight or partial internal melting, suddenly lose stability through a triggering effect, such as a gust of wind.

Soil-slide scars characterize steep slopes on smooth rock, where rootholds of plants tend to be insecure. (Adirondack Mts, N.Y.)

MUDFLOWS are movements of water-saturated clayey materials down streambeds. Mudflows—some fast, some slow—develop during heavy rains on fine, dry rock waste along steep slopes lacking vegetation. The porridge-like mass flows down to gentler slopes, where it comes to rest at channel margins. Thick deposits of volcanic ash on mountain slopes or of clay and silt on desert hillsides are likely materials for a mudflow. Lack of vegetation allows the fine material to be swept into gullies, where torrents churn it to make mud. These flows often block roads in desert country, and in volcanic regions they can quickly bury entire villages—as they did Herculaneum in Italy during the great eruption of Vesuvius in A.D. 79.

SOIL CREEP (DRIFT) is the very slow downslope movement of soil fragments at rates that diminish with depth. It results from disturbances such as ice-prying, impacts of raindrops, alternate wetting and drying, root growth, burrowing of animals, or subsidence of the

Mudflow on volcanic cone resulted when water saturated hot volcanic ash. Rise in foreground is edge of flow. (Mt. Lassen, Calif.)

Trees curved because they grow upward while mantle creep tends to tilt them downslope

Rock fragments creeping down by slight motions triggered by disturbances of mantle

ground. Fragments moved come to rest at a lower level than before. Thus when soil on a slope freezes, it is pushed out perpendicular to the slope, but when it thaws it is let down vertically, so that there is a net lowering with each freeze-thaw. Beneath the soil layer, upper parts of rock masses with closely spaced vertical joints or bedding planes may be bent downslope by ice-prying and the drag of soil creep.

Creep tends to be fastest at the surface, where disturbance is more likely and there is least stabilizing friction. Evidences of creep include downslope tilting of fences, grave markers, and telephone poles; also breached retaining walls and trees that curve out from the slope before turning up.

SUBSIDENCE, or sinking of ground, results from compaction or from collapse of cave roofs (p. 134).

Tops of basalt columns tilted downslope (to left) by creep: Close jointing makes such rock more vulnerable to distortion. (New Jersey)

SOLIFLUCTION is the movement, usually very slow, of a mass of rock waste over frozen ground. The slope may be only one or two degrees. Flows occur on alpine and arctic terrains where rock waste gets saturated with meltwater the frozen ground cannot absorb. A flow's edge may have lobes *(earth runs)*. Sometimes solifluction "stretches" stone rings (p. 31) into *garlands*. Solifluction was common around melting Pleistocene glaciers, and soliflucted material may be seen today on temperate lands that were glaciated.

ROCK GLACIERS are slow-moving accumulations of large rock fragments lubricated with water and ice. The edges of these masses are scalloped or lobed. As the ice melts and refreezes, over and over, the blocks of rock shift and the entire mass slowly creeps downslope like a glacier. Rock glaciers can exist as long as there is enough thawing to let the blocks shift but not enough to melt all the ice separating them. These features are seen on arctic and alpine terrains today. In the Rockies and New Hampshire's White Mountains, remains of Pleistocene rock glaciers are noticed as lobed masses of boulders on valley bottoms.

Solifluction mass showing lobes, or "earth runs" (Alaska)

Rock glacier (left) draping Wheeler Peak, Nev.

Gullies in roadcut demonstrate headward erosion in clay by streams.

SCENERY SHAPED BY STREAMS

Among all the land-shaping agents running water is by far the most effective. Streams, mainly, have cut the systems of valleys that drain the continents. Streams are the architects of flood plains, deltas, and other common deposit features. Not only full-sized rivers, but the countless small, temporary rivulets that result from every thaw and rain, work to shape the crust. The relief on landscapes generally, with some notable exceptions, has been shaped largely by streams.

RUNNING WATER originates as water evaporated from land and sea, lakes and rivers, then carried off by wind and eventually precipitated as rain or snow. Part of what falls on land becomes *runoff;* this flows over the ground until it reaches a stream, lake, or sea. Other precipitation becomes *ground water;* it percolates through soil and through joints and interconnected cavities in bedrock.

41

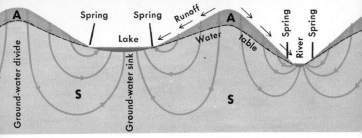

Ground-water circulation: Water from rainfall percolates through zone of aeration (A), where rock pores contain some air, to zone of saturation (S), where pores are full of water. Top surface of saturation zone is water table, which fluctuates with rainfall. Ground water follows curved paths in seeking level.

GROUND WATER filters down to where the soil or rock is already saturated. The surface of this zone is the *water table*. Its depth varies locally, being usually nearer the ground surface in humid regions than in deserts, in wet seasons than in dry seasons, and in valley bottoms than in highlands. Water tables tend to be very irregular surfaces, especially in distorted bedrock layers, but generally any valley bottom below the water table will have a stream, fed by ground water in the form of *springs*. Any basin with a floor below the water table will probably contain a *lake*, likewise fed by springs. Excess water leaves the basin by underground passages or, more commonly, flows over the lowest point on its edge, the *outlet*.

PERMANENT SURFACE STREAMS originate where water supply is steady, e.g. from melting on a mountain, from a depression that acts as a catch basin for runoff, or from a spring on a slope. On old lands, e.g. worn mountains, streams follow valleys already cut by long-continued stream activity. On relatively new lands, e.g. coastal plains or volcanic cones just built up from the sea bottom, running water is guided by existing depressions, which it gradually shapes into valleys.

STREAM CHANNELS The beds of streams—the surfaces along which the water flows—are *channels*. Shaped by the stream, they may zigzag or wind, according to *discharge* (volume and velocity) and *load* (sediments being transported). Channel banks may adjoin valley walls (p. 51) or be separated from them by a floodplain (p. 98).

Channels contain sediments resulting from weathering, mass wasting, and stream erosion. This waste is called *load* in a presently used channel, and *alluvium* when left exposed by shifting of the channel. Sediment from downcutting is mainly sand and silt, but fragments mass-wasted from valley walls may be of any size—boulders, cobbles, sand, mud. Finer sediments such as silt and mud tend to stay suspended in moving water. Coarser sediments, known as *bedload,* move by intermittent jumping, bouncing, rolling, and scraping. All this causes abrasion, which deepens the channel and rounds the sediments while reducing their size.

Erosion, transportation, and deposition: In a streambed, bare rock at **A** is exposed to erosion by abrasion and solution. At **B**, alluvial erosion occurs as one pebble is removed and transported to **C** but not replaced by a pebble from **A**. At **C**, transportation without erosion occurs as one pebble is moved to **D** and is replaced by one from **B**. Deposition occurs at **D** as a pebble there is covered by a pebble transported from **C**.

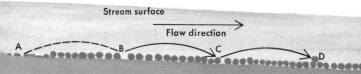

Stream surface

Flow direction

A   B   C   D

Rock

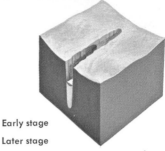

Early stage

Later stage

Valley development with
substantial mass wasting

Valley development with
little mass wasting

VALLEY DEVELOPMENT

VALLEYS ARE CUT primarily by streams eroding their channels. Erosion occurs by solution, by impact and suction in loose material, and by abrasion of bedrock as sediments are pushed or dragged against it. The faster the stream, the greater its erosive power, provided it is not overburdened with sediments. As long as the erosion process dominates, the stream's channel is cut lower and lower, and the valley deepens. The valley widens as a result of sidecutting by the stream and mass wasting on the valley walls. Meanwhile the valley may be extended in the upstream direction by *headward erosion*—gullying, undermining by springs, and slumping at the head.

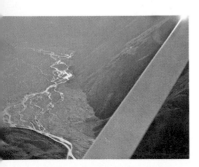

Air view of valley cut by a stream, then enlarged by a glacier: Present stream "braids" sediments on valley bottom. Source of stream is Murchison Glacier, New Zealand.

VALLEYS CHANGE in stages that may suggest a life cycle. In *youth* a valley has a swift stream that cuts down vigorously, making a deep ravine or gorge with waterfalls, plunge pools, rapids, and potholes (pp. 49-50). In *maturity* downcutting has considerably lowered the valley bottom; the stream is less vigorous. Downcutting has waned relative to sidecutting. Waterfalls and other irregularities have been eliminated by grading. The stream swings from side to side on a widened valley floor, depositing sediments to form small floodplains (p. 55). In *old age* the valley bottom is near base level—i.e. the limiting level of erosion, such as a large river on a broad plain, or the ocean. Downcutting has practically ceased, but sidecutting has made the valley still wider—perhaps miles wide. The stream meanders over a broad floodplain (p. 56) made by long ages of deposition. Changes affecting a stream's operations, e.g. uplift (p. 48), valley drowning (p. 48), or climate change, may interrupt a valley's evolution at any stage.

VALLEY PROFILES DURING LIFE CYCLE OF A STREAM

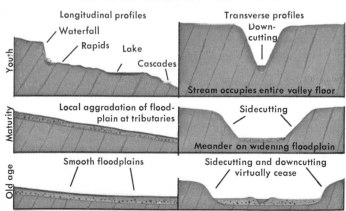

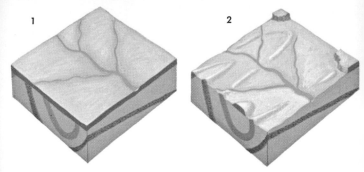

Formation of water gaps and a wind gap: (1) Streams flow on sediments covering eroded folds. (2) As mass wasting and erosion remove sediments, folds reappear. (3) Streams cutting across resistant belts

WATER GAPS are valley segments that cross the "grain" of ridges. Often such a gap is cut where a stream slices down through sediments covering eroded folds. When it has cut deep into the folds, it is called "superposed." Meanwhile other streams following weak belts may cut valleys parallel to the grain, so that the gap and the ridge it cuts through form simultaneously. Examples are seen in the Folded Appalachians and Rockies. Gaps may be cut also in structural sags, by headward erosion on flanks and crests of anticlines (p. 77), or as a stream holds its course across a rising fold.

Water gap was cut by superposed stream. (New River, Virginia)

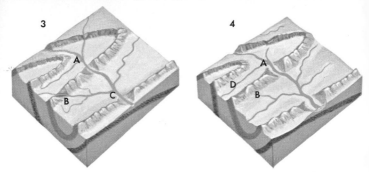

make water gaps at A,D,C, other streams by headward erosion along weak belts cause ridges to form. (4) By headward erosion, stream in weak belt intercepts transverse stream at D, making wind gap at B.

WIND GAPS are former water gaps, converted by stream capture. Capture occurs when one stream by headward erosion intersects another stream of lower gradient and diverts the water into its own channel. If the captured stream's channel was a water gap, it becomes a wind gap. Valleys flanking the ridge, at opposite ends of the gap, are cut deeper and deeper below the floor of the gap. Eventually the gap may appear as but a shallow notch in the ridge. Examples: Snickers, Ashby, and Manassas gaps in Virginia's Blue Ridge—water gaps converted into wind gaps by the Shenandoah River's headward erosion.

Wind gap was left high above valley. (New Market Gap, Virginia)

Meanders, or "goose necks," were cut deep after stream was rejuvenated. (San Juan River near Mexican Hat, Utah)

A NARROW INNER VALLEY in a wide one may indicate renewed downcutting by a stream that had ceased to cut. Renewed ability to cut, called *rejuvenation,* results from uplift or tilting of the land or from climatic change. The horizontal surface between the inner valley and the old valley wall is a remnant of the former valley floor. If it is bedrock, it is called a *strath terrace;* this is cut regardless of rock structure (compare with alluvial terraces, p. 55, and benches, p. 52). In recently uplifted regions, e.g. the Colorado and Appalachian plateaus, rejuvenation has been common.

ESTUARIES are seaward ends of valleys invaded by the sea as a result of a rise in sea level or subsidence of the coast. In these "drowned" valley mouths downcutting has ceased and sediments are accumulating. Examples: North America's Atlantic Coast from the Gulf of the St. Lawrence to Chesapeake Bay (p. 143), and European coasts from England to Spain.

EROSIONAL FEATURES OF VALLEYS

POTHOLES: deep, smoothly circular or elliptical depressions in bedrock made by abrasive sediment in eddies. In young valleys; up to yard deep and wide. Even small ones may take decades to form.

Potholes (Arizona)

WATERFALLS: typical of young valleys. Cliff with falls may be edge of rock mass more resist un mun rock below falls. Or cliff may result from faulting, glaciation (pp. 116, 153), or valley-blocking by landslide or lava flow. Close series of falls form cascade. Falls tend to be gradually eliminated by abrasion and solution at top of falls and undermining at base.

Waterfall in volcanic area may occur where streambed is blocked by lava flow or disrupted by earth movement. (Gullfoss, Iceland)

Falls of Niagara type forms where stream flows over edge of resistant layer into depression made by stream in less resistant rock (brown in diagram).

Rapids occur where Potomac River descends over strong metamorphic rock near east edge of Virginia Piedmont before traversing weak sedimentary rock of coastal plain. (Great Falls, Va.)

RAPIDS: areas of swift, dashing water ("white water") around rocks projecting from streambed. Rocks may be blocks weathered off valley walls, or may be edges of resistant strata—perhaps last vestiges of a waterfall. Rapids are common, as are waterfalls, in areas of recent volcanic activity or faulting. Rapids tend to disappear as valley bottom becomes graded by long erosion.

PLUNGE POOLS: bodies of water at base of waterfalls in basins made by erosive action of falls. May extend to undermined region, permitting observer to walk behind falls.

Plunge pool was formed by prolonged abrasive action of sediments in churning water at base of falls. (Cherokee Nat. For., Tenn.)

V-SHAPED VALLEY CROSS SECTIONS: formed where mass wasting on walls keeps pace with stream erosion. Loose mass-wasted material makes slope of 35° (angle of repose) or less. This if combined with cliff makes a "complex" valley wall.

V-valley cut by highland stream (Wasatch Mts, Utah)

UNDERCUT SLOPE: steep slope or cliff on outside of river bend. Sidecutting by stream as it rounds curve tends to produce unstable overhangs. These fall, leaving steep wall.

SLIP-OFF SLOPE: gentler slope on inside of river bend, opposite undercut slope. Sidecutting on inside of bend is minimal; hence the gentler slope. Often fringed by point bar (p. 55).

Undercut and slipoff slopes (Ardèche near Chames, France)

VERTICAL WALLS: developed where downcutting occurs without much mass wasting on valley walls or sidecutting by stream. Relatively rapid downcutting may be favored by vertical joint, fault, or weak zone (e.g. a dike, p. 74). Chasm or gorge may result. Where channel is relatively straight, stream makes slot-like cut. At bends in strong rock where weathering and mass wasting do not keep pace with sidecutting, overhangs may develop, forming alcoves or niches (p. 9).

Vertical-walled valley was cut by stream following master joint. Edges of more resistant sandstone layers make "ribs." (Ausable Chasm, N.Y.)

Valley walls in profile here are formed by dip slopes (surfaces parallel to dipping strata) and scarp slopes (surfaces at angle to strata). Steeper walls are scarp slopes. (Whetstone Mts, Ariz.)

WALLS OF UNEQUAL SLOPE: developed where effects of weathering, mass wasting, and stream erosion differ on opposite sides of valley. Thus in temperate regions of northern hemisphere an east-west valley in homogeneous sedimentary rock is less susceptible to weathering and mass wasting on its north-facing wall than on

south-facing wall, where thaw-freeze is more frequent. In another situation, selective erosion by a stream following weak belt in tilted sedimentary strata may develop an asymmetric valley, the gentler slope being with dip of the strata.

RIBS: edges of relatively resistant, nearly horizontal strata projecting from valley walls. Generally alternate with recessed edges of weaker strata. Characteristic of layered rocks, especially sedimentaries (see photo at left and at bottom of p. 51).

BENCHES: rock platforms extending from base of scarp, or cliff, on valley wall toward stream channel cut to lower level. Usually are parts of strata resistant to mass wasting but succumbing to stream erosion. Material from above may wash across bench and broaden it. Bench typically on single strong layer overlain by weak layer.

Benches on resistant layers of shale (Watkins Glen, N.Y.)

Natural bridge (Pont d'Arc near Vallon, France)

Buttresses of sandstone (Palo Duro Canyon, Tex.)

NATURAL BRIDGES: rock arches of various origins. One kind is made by a stream cutting through a divide; thus a meander cut off (p. 56) made the Pont d'Arc of the Ardêche in France. Others are remnants of stream-made tunnels, e.g. Natural Bridge, Va., and Rainbow Bridge, Utah. Some result from marine erosion (p. 148) and collapse of parts of cave roofs.

PINNACLES AND TOWERS: usually result from predominantly vertical jointing. Joints define the forms; weathering and

erosion isolate them from each other and from valley wall. Some are hard cores of volcanic material (e.g. volcanic necks) from which weaker surrounding material has been eroded away. May occur as erosional remnants on plains and plateaus. Occasionally are weak material protected by resistant cap (p. 121).

BUTTRESSES: nearly rectangular rock masses projecting from valley walls; may resemble buttresses in Gothic churches. Often in sedimentary rocks with wide rectangular jointing.

Vertical joints guided sculpturing of pillars. (Meteora, Greece)

DEPOSIT FEATURES

A stream's ability to transport solid particles depends on its volume and velocity and on the density, quantity, and caliber of the particles. If material delivered to the stream exceeds the transporting ability, some must be deposited. Coarse material *(bedload)* is generally deposited where turbulent flow is no longer competent to move it along the bed. Finer material is held in suspension until loss of turbulence or flocculation (sticking together of grains) causes settling.

These changes may occur all along a streambed, so that deposits may be found anywhere at times of moderate flow. Deposits may range from large chunks of slide rock to fine silt and clay. The various sizes are sorted during transportation. While material is within reach of the stream it is subject to reworking. It becomes alluvium when isolated by a shift of the channel.

ALLUVIAL CONES AND FANS: masses of sediment deposited at foot of mountain ravines where streams reach valley floor abruptly. Sediments are sorted, grading from coarse to fine at bottom of slope, according to change of gradient. Alluvial cones, seen at bottom of steeper slopes, differ from talus cones, in which material is not sorted. Alluvial fans, formed at foot of gentler slopes, may grow much larger than cones, becoming plains (pp. 98-99). Spectacular in deserts.

BRAIDING: division of stream into channels separated by elongated islands of sediment (p. 44). Occurs if stream has overload of coarse material, slows, and can spread laterally when channel becomes choked with sediment. Common where streams issue from valley glaciers into abandoned glacial troughs (p. 120), and in deserts and semi-arid regions.

Alluvial cones fringing a steep scarp (Echo Cliffs, Ariz.)

Alluvial terrace is on sand deposited by stream from melting Pleistocene glacier. (Vermont)

ALLUVIAL TERRACES: level surfaces on deposits extending out from valley wall. Terrace represents former level of stream deposition. Change in conditions of flow caused stream to regrade itself (resmooth its bed) at lower level, thus leaving the former deposition level considerably above the stream. Alluvial terraces are subject to dissection by tributaries and by the mainstream. Common in valleys whose streams were formerly swollen with rock waste and with meltwaters from shrinking glaciers.

FLOODPLAINS: nearly horizontal surfaces on fine-grained deposits left by floods. The floodwaters are overflow from a curved channel no longer being cut down. Elevation of channel floor varies according to flood conditions. Making of floodplain starts with deposition of sediment as *point bars* at base of slipoff slopes while channel is shifting laterally around bends. For specific floodplain features see diagrams at right; also next page.

FLOODPLAIN DEVELOPMENT

(1) Stream begins meandering; point bar initiates floodplain. (2) Meander loop migrates downstream, trimming spurs and widening point bars. (3) Fully developed floodplain has spurs trimmed to width of the meander belt.

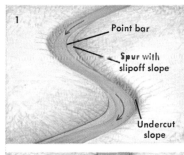

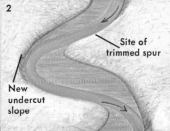

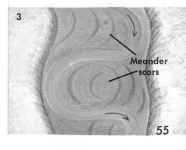

55

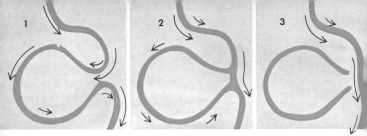

Formation of oxbow lake: Arrows show direction of streamflow; the longer the arrow, the faster the flow. (1) Almost complete meander circle develops. (2) High water floods across neck, making cutoff. (3) Regrading of cutoff seals ends of loop, forming oxbow lake.

MEANDERS: smooth, looplike curves developed by continued cutting and filling in channel. Meanders migrate downstream; meanwhile spurs are trimmed and eliminated (p. 55).

CUTOFFS: short channels across necks of meander curves. As curve nears full circle, neck narrows. Flood across neck may erode new, shorter channel.

OXBOW LAKES: often formed in cut-off meander loops (see diagram above). The lakes are fed by seepage.

MEANDER SCARS: meanders, or parts of them, isolated by cutoffs and now filled with sediment. Best seen from air. Scars in series show how stream repeats cycle of establishing a meandering course, cutting off and filling loops, then making new channel loops downstream from old ones.

NATURAL LEVEES: arcuate or curved, round-backed ridges flanking meanders, 10-12 ft. above floodplain. Formed as floodwaters, overflowing channel, deposit most of load close to it. Levees nearly parallel channel. Stream tends to build up its bed between levees; thus stream level is higher at flood time and levee breaks are serious.

Floodplain exhibits meanders, cutoffs, oxbow lakes, and meander scars. (Wisconsin)

Delta of arcuate type results from deposition by closely spaced distributaries. Light areas are shore deposits. (Cook Inlet, Alaska)

DELTAS: accumulations of sediments deposited where streams empty into lakes or oceans. Shape depends mainly on shoreline contours and currents (p. 142). Some arcuate deltas, e.g. Nile Delta, suggest Greek letter (△). Mississippi Delta is bird-foot (digitate) type. Deltas in estuaries or lakes are long and narrow. With each flood a delta receives more sediments, which are spread by small branches (*distributaries*) from mainstream. Deltas fertile where sediments are rich in organic wastes.

ARCUATE DELTA   BIRDFOOT DELTA   CUSPATE DELTA

Mediterranean Sea

Nile Delta

EGYPT — Nile River

Mississippi River and Mississippi Delta

Gulf of Mexico

ITALY — Tiber River

Tiber Delta

Tyrrhenian Sea

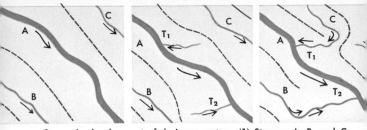

Stages in development of drainage system: (1) Streams A, B, and C are separate; (2) tributaries T_1 and T_2 develop from stream A in largest drainage area; (3) by headward erosion T_1 captures stream C, T_2 captures stream B. Thus smaller streams are diverted into larger ones. Dashed lines show how divides shift.

DRAINAGE SYSTEMS are integrated groups of valleys by which running water on land moves to lower and lower levels. As a new land surface forms (e.g. a coastal plain), valleys will begin developing on its slopes. Mostly by headward erosion they link with other valleys, forming a system of drainage lines. Patterns in a well-developed system depend on underlying rock structures (pp. 79, 86). Drainage can be traced on contour maps and well observed from the air.

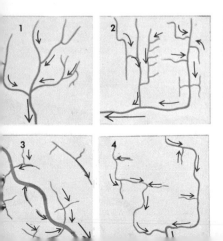

COMMON DRAINAGE PATTERNS (1) Dendritic: little linear control. (2) Trellis: linear control by folds. (3) Annular: tributaries ringlike from influence of domes and basins. (4) Rectangular: controlled by fault or joint lines. Also common is the parallel pattern (not shown here), with nearly parallel stream courses controlled by tilted rock layer.

DISSECTION OF THE LAND proceeds by stages. For example, suppose a wide, flat-surfaced, homogeneous rock mass is exposed rapidly by uplift of a sea floor. Streamwork, with weathering and mass wasting, eventually lowers the land to sea level. In this long process the first stage—relatively very short—is that of *youthful dissection,* in which a few widely spaced streams cut V-valleys. Divides between streams are broad, relief is increasing, the land is high. In the stage of *mature dissection* the stream system is well filled out with tributaries, and the larger streams are fully mature. Divides are narrow, relief at maximum, and the original level of the region is preserved on the hilltops. In *old age* the larger streams are old and many of the tributaries mature; divides are widely spaced, slopes gentler, altitudes lower. The region may now be a *peneplain* (Latin *pene,* "almost"): a rolling surface eroded low except for scattered hills (*monadnocks*). Such evolution may be modified in various ways (p. 45).

Stream dissection cycle:
evolution of a landscape.

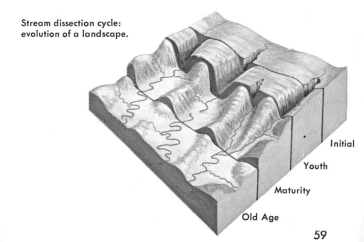

Initial

Youth

Maturity

Old Age

59

Steep-sided volcanic cone was built of alternating, steeply sloping layers of flow lava and blown-out materials. A common continental type. (Mt. Shishaldin, Aleutian Is., Alaska)

LANDFORMS ON IGNEOUS ROCKS

Igneous activity has occurred intermittently since Earth's earliest days. It involves generation of heat at depth, formation of magma, and the rise of magma through crustal fractures. Intense igneous activity typically occurs in zones of concentrated high stress, i.e. along rifts (p. 96) and in young mountain ranges.

Sites of igneous activity change from age to age. Most parts of the world were once volcanic. The first lands were formed probably by piling up of lavas from vents in sea bottoms. This process is still making oceanic islands—i.e. islands not resulting from a change in sea level or coastal erosion. Much of the crust consists of igneous rock formed in the remote past, and of rock waste from erosion of ancient volcanic highlands. Where volcanism has occurred in recent geologic time, the resulting landforms still exist—volcanic cones, lava plains, and lesser features. Volcanic landforms produced in the distant geologic past have generally eroded away except for their roots.

Several great mountain ranges have been built up by volcanic action. These mountains and their relationship to mountains of other kinds are described on pp. 95-96.

Igneous rocks are classified as (1) *volcanic,* or *extrusive* (formed on the crust); and (2) *plutonic,* or *intrusive* (formed within the crust). Each rock category yields a characteristic array of landforms. These forms are determined by conditions of eruption or intrusion and by gradation.

Volcanic regions: Areas active recently, including about 500 volcanoes, are shown in red on map. (Not shown is Antarctica, which also has active volcanoes.) Numbers on map designate parts of volcanic island arcs, which follow rifts in the crust: 1 Heard I.; 2 New Zealand; 3 Indonesia; 4 Philippine Is.; 5 Japan; 6 Kurile Is.; 7 Aleutian Is.; 8 Hawaii; 9 Iceland; 10 Azores; 11 Canary Is.; 12 St. Paul Rock.

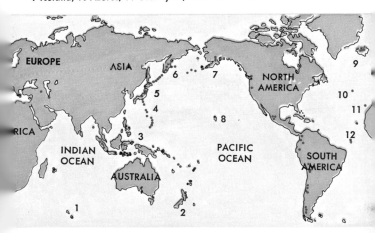

Craters on a shield volcano follow a rift. (Mauna Loa, Hawaii)

SCENERY ON EXTRUSIVE ROCK

VOLCANIC CONES are built up as lava is erupted from pipes. A pipe is a conduit through the crust, often where fractures intersect. Periodically magma from a deep reservoir rises in the pipe and flows out onto the surface or is blasted out; thus lava accumulates. A crater is made in the cone by eruptions, by undermining and collapse of material around the vent, and by sinking of lava in the pipe after eruptions.

Cones with the steepest slopes (up to 40 degrees) are made of volcanic cinders (p. 66), erupted in relatively mild explosions. These *cinder cones*, usually several hundred feet high, frequently have well-developed, symmetrical craters (p. 65). They absorb rain and meltwater because of their loose construction and high permeability; thus gullying may be minimal. Extinct cones, well preserved, are seen in recently active areas, e.g. Arizona and New Mexico, the Northwest, and the Auvergne area in France.

Cones with the gentlest slopes are made by eruptions of basaltic lava (pp. 23, 67) at high temperatures (up to 2,200°F.) with little entrapped gas. Such lava, very fluid, emerges quietly and may flow at 5 to 25 m.p.h., reaching 30 to 40 mi. from the vent. Repeated flows build up a broad, shield-shaped cone, or *shield volcano*. Slopes approximate 5 degrees near the summit, where the lava is hottest and most fluid, and increase gradually to 12 degrees near the base, where the lava is cool and tends to pile up. Shield forms characterize mid-ocean volcanoes (Hawaii, Iceland).

Other volcanoes, particularly those on or near large land masses, tend to produce felsitic (rhyolitic) and andesitic lavas (p. 23). These, more viscous than basaltic lavas, cannot flow as far. They tend to block the volcanic vent, causing explosions. Alternating layers of flow lavas and blown-out fragments build up a *composite cone,* or *strato-volcano*. This has relatively steep slopes because the flows and most of the blown-out material come to rest near the vent. Well-known composite cones include Mts. Rainier and Hood in the Cascade Mountains, the San Francisco Peaks in Arizona, and the great Italian volcanoes Vesuvius and Etna.

STRUCTURE OF A TYPICAL STRATO VOLCANO

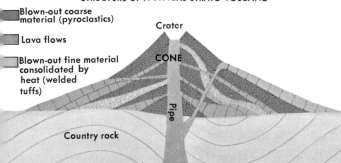

Blown-out coarse material (pyroclastics)

Lava flows

Blown-out fine material consolidated by heat (welded tuffs)

Crater

CONE

Pipe

Country rock

Extinct cones rising above quiet countryside (Auvergne, France)

Depression due to subsidence in flow (Craters of Moon, Idaho)

FEATURES ASSOCIATED WITH CONES

PARASITIC CONES: built up on sides of larger cones by eruptions through subsidiary vents.

SPATTER CONES: formed by ejection of lava froth through small vents on or near cones.

PIT CRATERS: depressions on floor of large crater due to local withdrawal of lava.

PLUG DOMES: masses of viscous lava that well up from a crater or other volcanic vent and solidify into rounded form. May develop on flanks of large cone, or in crater, or as separate feature over hidden volcanic vent, e.g. Mt. Lassen, Calif., and Puy Grand Sarcoui, Auvergne (France).

SPINES: towerlike masses of solid lava pushed up out of cones by renewed activity. May rise hundreds of feet. Hardened lava flakes off. Weight of spine and internal pressure may burst cone, e.g. Mt. Pelée (p. 68).

Rhyolite plug dome in pumice cone (Katmai Nat. Mon., Alaska)

Caldera of Mt. Mazama, formed by collapse of crater walls about 6,400 years ago, holds lake about 2,000 ft. deep. Recent cinder cone, Wizard Island, rises from caldera floor. (Crater Lake, Oreg.)

CALDERAS are craters much enlarged by explosions and by collapse of the walls due to undermining. The floor may have small domes (*tumuli*) produced by pressure from below. After a period of dormancy new eruptions may build a new cone on the floor. Some calderas have several cones nested; e.g. Mt. Vesuvius. Crater Lake, Oreg., occupies the 6-mi.-wide caldera of a Pleistocene volcano, Mt. Mazama, perhaps 12,000 ft. high before its top was destroyed by explosions and collapse. Wizard Island, a recent cone, rises above the water. Valles Caldera (diameter over 13 mi.) in New Mexico is the crater of extinct Jemez Volcano, from which an estimated 50 cu. mi. of rhyolite fragments were blown out. Hundreds of square miles around Crater Lake and Valles Caldera are covered with volcanic ash and larger lava fragments yards deep.

Volcanic bombs (Idaho)   Scoria (New Mexico)

FORMS OF ERUPTED MATERIAL

Erupted materials are of two general kinds: *pyroclastics,* which are blown out of volcanoes as fragments of various sizes and shapes; and *flow lavas,* which come from pipes or fissures and may flow many miles before solidifying. **Pyroclastics** include the following:

VOLCANIC BLOCKS: fragments of rock torn from walls of volcanic pipe during eruptions.

VOLCANIC BOMBS: large blobs of lava blasted into the air, then cooled and solidified in falling; often spindle-shaped.

SCORIA: usually, dark, clinkery lava fragments with many holes left by escaping gases; usually basaltic; heavier than pumice.

CINDERS: fragments 1 to 3 in. in diameter; full of gas holes; may be partly crystalline, partly glassy.

Lava flow covering tuff exposed in roadcut (Oregon)

LAPILLI (Italian, "little stones"), *Pelé's tears,* or *Apache tears:* 1/7 to 1 inch in diameter.

VOLCANIC ASH: particles 1/100 to 1/7 in. across.

VOLCANIC DUST: particles smaller than volcanic ash.

VOLCANIC TUFF: compacted volcanic ash or dust, sometimes welded by volcanic heat; often dissected by streams to form pillars and cliffs (p. 5).

TUFF BRECCIA: mixture of tuff with angular lava fragments.

Field of recent bombs and cinders (Craters of Moon, Idaho)

Pahoehoe (foreground) and aa
(Craters of Moon, Idaho)

Pressure ridge on basaltic flow
(Grants, New Mexico)

BASALTIC LAVA FLOWS from cones or fissures may spread over large areas, forming plains and plateaus. The solidified lava is the rock basalt. Flows are well represented in Hawaii (pp. 62, 142), the Columbia Plateau (pp. 107-108), and northern parts of New Mexico and Arizona.

PAHOEHOE (puh-HOI-hoi): from very liquid lava; often cools as ropy-looking coils; common Hawaiian form.

PILLOW LAVA: pillow-like lava, solidified underwater.

AA (AH-ah): rough, broken, blocky crust formed as flow becomes viscous and hardened while being pushed by still-fluid lava behind it.

LAVA CAVERNS AND TUBES: formed as liquid lava flows out from within an enclosing crust.

LAVA STALACTITES: dripping lava solidified; usually in cavern or tube.

BASALT COLUMNS: made by jointing of compact lava as it cools and contracts. Cross sections often six-sided; may be five-, four-, or three-sided.

PRESSURE RIDGES: bulges on lava due to flow pressure.

SQUEEZE-UPS: lava masses squeezed up through breaks in crust of flow. May be linear, bulbous, or irregular.

Lava tunnel a mile long, 50 ft. high (Lava Caves, Oreg.)

Pillow lava of ancient flow in quarry wall (New Jersey)

Distorted flow lines in rhyolite are exposed in cliff. (Idaho)

Canyon was cut in mineralized rhyolite pyroclastics. (Yellowstone Nat. Pk., Wyo.)

FELSITIC LAVA FLOWS are viscous. If the lava contains little gas, it may extrude from a central vent as a *plug dome* or *spine* (p. 64). If gas pressure builds up beneath a dome or spine that is stronger than adjacent parts of the cone, the cone may burst, spraying incandescent solids, liquids, and gases widely. The glowing cloud can quickly cover many square miles. A cloud from Mt. Pelée devastated St. Pierre, Martinique, in 1902, killing 30,000. Such clouds may have made some of Yellowstone's extensive rhyolitic deposits.

Felsitic flows cool and solidify to form light-colored felsite (p. 23), including rhyolite. Very frothy lava solidifies into *pumice,* a whitish or pale-gray rock often light enough to float. Some gasless felsitic lavas cool fast enough to form a non-crystalline rock—*obsidian,* glassy and black.

Pumice

Obsidian

Erosion made pinnacles where masses of weak volcanic material were capped by resistant rock fragments. (Oregon)

Straight, deep vertical-walled valleys suggest characteristic rapid erosion of basalt under humid climate. (Hawaii)

EROSION OF LAVA

Recent flow lavas and pyroclastics vary in resistance to erosion. Basaltic lavas on humid South Pacific islands have been deeply and widely eroded, so that many volcanic cones are hardly recognizable. In temperate and arctic zones, cones and flows of felsitic lavas resist erosion strongly. Highly permeable pyroclastics in cinder cones absorb water and thus undergo little stream erosion, but they are extremely vulnerable to wave attack, as in the island built up by Iceland's Surtsey Volcano. Less permeable pyroclastics tend to gully quickly. During eruptions, rain from condensed steam cuts deep, straight gullies on the flanks of a cone. Hot mud pours down these gullies and comes to rest on lower slopes (p. 38). Where lavas vary in resistance, e.g. where thick pyroclastics alternate with thin flows, a myriad of grotesque erosion forms may result.

Repeated eruptions by large volcanoes bury thousands of square miles under masses of breccia, cinders, and tuff. This is highly erosible, often yielding badland topography with a wide variety of relief (above left, and pp. 5, 23).

MINOR VOLCANIC FEATURES

After volcanoes become dormant or extinct, residual heat in the crust may produce certain minor features:

FUMAROLES: fissures producing volcanic steam mostly. Fissures producing much sulfurous acid, from which sulfur may be deposited, are *solfataras*. Water vapor comes partly from air, partly from volcanic source; sulfurous products apparently come from magma chamber. "Bumpass Hell" at Mt. Lassen, Calif., has fumaroles and solfataras. Mt. Etna, Italy, has over 500 fumaroles. Greatest display of fumaroles was in Alaska, in Valley of 10,000 Smokes, after great eruption of Mt. Katmai, 1912.

HOT SPRINGS: ground water that percolated to hot zone through joints and tubes, then rose by convection, forming pools at surface. Springs develop after most acid gases dissipate. Yellowstone's springs may have water from 8,000 ft. down; temperatures average 170°—cool enough for algae, which with minerals color water. Idaho has hotter springs, probably from magma. Arkansas' hot springs are about 142°.

MUDPOTS: hot springs full of mineral mud, kept churning and bubbling by escaping vapors.

Fumaroles (Mt. Lassen, Calif.)

Fumarole cores that resisted erosion (Crater Lake, Oreg.)

Hot spring under cloud of steam. (Yellowstone Nat. Pk.)

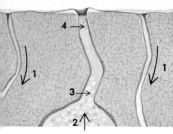

Conditions just before geyser eruption: (1) ground water descending to zone of hot rock; (2) superheated water flashing into steam; (3) steam rising in pipe; (4) water at lower temperature being forced up toward vent.

Geyser needs narrow vent, high steam pressure. (Hveragerdi, Iceland)

GEYSERS: hot springs that periodically spew steam and hot water, then subside. Require rare conditions; occur in few areas, notably Yellowstone Park and The Geysers, Calif.; Iceland (site of historic Geysir); New Zealand. Some, like Old Faithful in Yellowstone, erupt fairly regularly.

Eruptions result from the superheating of ground water trapped in underground passages connected with main pipe of geyser. When temperature of the trapped water gets high enough, the water flashes into steam, expelling other water out of the vent violently. Thus pressure on water below eases and more water vaporizes. Process ceases as ejected hot water is replaced by cooler water entering underground.

RIMS, DOMES, TERRACES: masses of travertine (p. 138) and silica (p. 20) deposited by hot springs and geysers. May survive as landforms long after activity has ceased. In Yellowstone much silica is still being deposited; also, travertine in area of Mammoth Hot Springs.

Travertine terraces are colored by impurities. (Mammoth Hot Springs, Yellowstone Nat. Pk.)

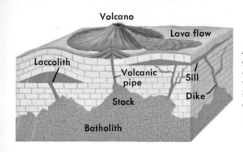

An active volcanic region: Volcanic activity always is associated with the formation of igneous structures underground.

Labels on figure: Volcano, Lava flow, Laccolith, Volcanic pipe, Sill, Dike, Stock, Batholith

SCENERY ON INTRUSIVE IGNEOUS ROCK

Plutonic activity, involving formation of intrusive igneous rock, generally accompanies volcanic activity. Intrusive rock masses vary widely in size, chemical composition (p. 22), and physical form. They become landforms when exposed and shaped by erosion.

BATHOLITHS (Greek, "depth" + "stone"): huge bodies of granite, 40 mi. or more in extent, formed deep in crust long ago. Uplift, due in part to relative lightness of granite, results in erosion of overlying material and eventual exposure of batholith's top. Granite forms long-lasting highlands, sometimes ringed with stumps of strata that formerly covered batholith. Part of California's Sierra Nevada Range is batholithic. The Clearwater, Salmon, Sawtooth, and Coeur d'Alene ranges have been cut in the giant Idaho batholith.

Batholith was cut into peaks by erosion. (Adirondack Mts, N.Y.)

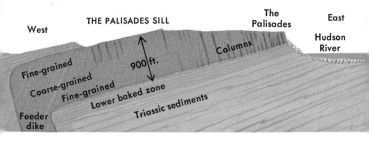

THE PALISADES SILL

West · The Palisades · East · Hudson River

Fine-grained
Coarse-grained
Fine-grained
900 ft.
Columns
Lower baked zone
Feeder dike
Triassic sediments

STOCKS: smaller than batholiths; develop near large ranges; may feed magma to volcanoes or inject it into crust. Stocks exposed by erosion become highlands if relatively resistant; e.g. Crazy and Judith Mts. of Montana, Henry Mts. of Utah.

LACCOLITHS (Greek, "lentil" or "lens" + "stone"): intrusive masses that caused doming at time of intrusion. Apparently fed from stocks. Mostly in sedimentary strata east and west of Rockies. Cluster northwest of Black Hills includes Crow and Elkhorn peaks, Crook Mountain, and Little Sun Dance (Green) Mountain (which still has sedimentary cover). Other laccoliths: Highwood Mts., Mont.; Navajo Mt, south Utah; Packsaddle Mt, west Texas.

SILLS: sheet- or slab-like masses intruded between strata. Most are basaltic. Eroded edges of sills in nearly vertical strata form ridges or valleys. In gently dipping strata, sill edge exposed over weaker rock may form cliff. Hudson Palisades of

N.Y.-N.J., 20 mi. long, are mostly edge of westward-dipping, 900-ft.-thick sill injected between relatively weak sandstones and shales. Great Whin Sill, north England, forms cliffs along 40-mi. front between Farne Is. and Pennine Hills.

Laccolith stripped of covering strata (Carrizo Peak, N. Mex.)

Great Whin Sill, northern England (Near Twice Brewed)

Dike ridges are common where igneous rock invaded relatively weak host rock. (Colorado)

Dike of dark basic rock intrudes this granite. Note sheeting in granite. (Mt. Desert, Maine)

Volcanic necks made good sites for castles and churches. (Chapel St. Michel d'Aiguilhe, Le Puy, France)

DIKES: wall-like masses that cross the grain or strata of host rock (rock into which they are intruded). May be basaltic, andesitic, or felsitic; straight, curved, or circular (ring dikes). Groups of dikes sometimes radiate from magma reservoir, usually a stock (p. 73). When eroded, vertical or nearly vertical dikes form ridges if more resistant than host rock, valleys if less resistant. Such ridges are common in south-central Colorado, where host rocks are sedimentary. In regions of granitic and metamorphic rock, basaltic dikes are usually relatively weak, hence eroded to form valleys, e.g. The Flume in Franconia Notch, N.H., and countless similar narrow valleys in Rockies and older Appalachians. Waterfalls or rapids often found where streams traverse dike rock and host rock of sharply differing resistance.

VOLCANIC NECKS: masses of lava, mixed with broken bedrock, that solidify in volcano's pipe as activity ceases. If pipe filling is more resistant than cone or its foundation, it remains as tower or knob after erosion has destroyed cone. Many necks dot Southwest and parts of Colorado Plateau. Some very old necks in Appalachians. Examples in British Isles (esp. Ireland and Scotland), central France (Auvergne), W. Germany.

Syncline in limestone dominates scene. (Wasatch Mts, Utah)

FOLD AND FAULT PATTERNS

Everywhere, especially in the vicinity of young mountain systems, the crust is responding to forces acting within it and on it. These forces are compressive (squeezing), tensile (stretching), torsional (twisting), and shearing (acting in opposite directions in the same plane). They cause folding (distortion) and faulting (breakage with dislocation), usually in very small increments over enormous time spans (pp. 17-18). Such events may produce relief directly, as when a crustal block is tilted to make a highland. Folding and faulting may produce landforms indirectly also, by continuing to guide erosion long after the original relief has eroded away. A common example is the production of relief by differential erosion of exposed roots of a deeply eroded mountain mass. Folds and faults have influenced most landscapes that are mainly erosional.

FOLDS AND FAULTS usually are partly concealed or eroded. They are seen mostly in roadcuts, quarries, and other wide bedrock exposures. Evidence of folding and faulting may be discerned in the tilt or curve of rock outcrops and profiles of hills and valleys. "Topo" maps (p. 154) help in tracing of folds and faults.

FOLDING

Rock can be distorted without crumbling. When forces beyond the elastic limits are slowly applied to rock under confining pressure for long periods, they induce permanent bends. In stratified rocks, layers may slide against one another like playing cards when the deck is bent. Or minerals may recrystallize under high pressure and align themselves parallel to the fold axis. In such ways many kinds of folds are produced, ranging from the microscopic to features miles wide and deep, and hundreds of miles long.

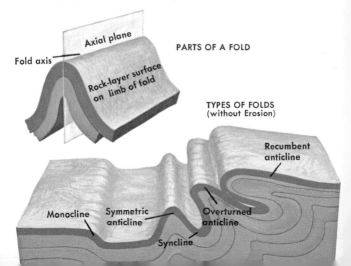

Axial plane

Fold axis

Rock-layer surface on limb of fold

PARTS OF A FOLD

TYPES OF FOLDS
(without Erosion)

Recumbent anticline

Monocline

Symmetric anticline

Overturned anticline

Syncline

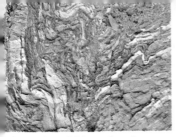

Intense folding of sedimentary layers (Glacier Nat. Pk., Mont.)

Anticline and syncline in beds of sandstone (New Jersey)

FOLD TYPES

WARPS: minor distortions due to bending and twisting. Occur in most layered rocks that were raised or lowered. Show usually as slight changes in slope and elevation over wide area.

MONOCLINES: local steepenings of dip in layered rock. Usually in near-horizontal sedimentary strata, e.g. Colorado Plateau, or in flanks of folds.

ANTICLINES: tent- or arch-shaped upfolds. Usually joined with synclines (below) in groups with straight or gently curving parallel axes. Typical fold-mountain structures (pp. 89-91).

SYNCLINES: trough- or V-shaped downfolds. Usually occur with anticlines (above).

OVERTURNED FOLDS, folds with one limb rotated more than 90 degrees from the original horizontal. Axial plane inclined; both limbs slope in same direction (pp. 92-93) (see p. 76).

RECUMBENT FOLDS: axial plane and limbs nearly horizontal, lying on their sides (p. 93).

PLUNGING FOLDS: axes inclined to horizontal. Fold plunging at both ends is *doubly plunging*. Folds of this type include domes (p. 80).

HOMOCLINES: masses of strata that dip in same direction. They are not, strictly speaking, due to folding, but may result where broad warping has been followed by faulting.

Folding in limestone, with faulting (Galliluro Mts, Ariz.)

Small monocline seen in limb of large fold (Belt Mts, Mont.)

Synclinal mountain (Mt. Kerkeslin, Jasper Nat. Pk., Canada)

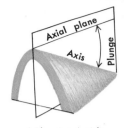

A Plunging Anticline

FOLDS INFLUENCE RELIEF by guiding erosion. Gradational agents begin to erode anticlines as these rise above sea level; thus we seldom see folds uneroded. Erosion may vary from a single water gap cut through the upper layers to the removal of several layers, exposing a full anticlinal ridge. More frequently, long erosion develops ridges on upturned edges of resistant strata and valleys in weak bands. How these controls operate is shown on p. 79. Groups of large folds form mountains (pp. 87-93).

Front Range of Rockies exhibits folds. (Jasper Nat. Pk., Canada)

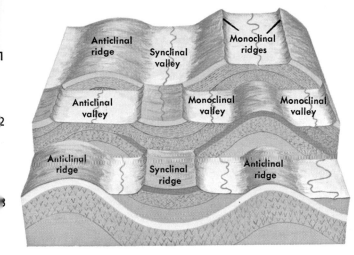

Erosion on folds: original surfaces (1) partly eroded, (2) eroded away except for syncline (topography is inverted), (3) all eroded away.

Where differential erosion, rather than the fold it-self, accounts for relief, anticlinal valleys and synclinal ridges are common, forming *inverted topography*.

ANTICLINAL RIDGES: Summits are crests of anticlines, usually broad and rounded. May be original warped surface slightly eroded, or lower resistant layer exposed by deep erosion. See facing page.

SYNCLINAL RIDGES: Summits are synclinal troughs, usually flat-topped; some have lengthwise hollow. Steep-sided, blunt-ended; usually left by deeper erosion of adjoining anticlines. See facing page.

ANTICLINAL VALLEYS: Follow axis of anticline. Usually steep-sided.

SYNCLINAL VALLEYS: Follow axis of syncline. Slopes often gentle.

MONOCLINAL (and HOMO-CLINAL) RIDGES AND VALLEYS: Formed by erosion of beds all dipping in same direction. Slopes usually unequally steep (p. 52). Sharp-backed, steep-sloped ridges are "hogbacks."

79

DOMES are upfolds with little elongation along the axis. They form ovals or near-circles. Topography may be inverted. Domes result mainly from vertical uplift; e.g. the Black Hills Dome, where old igneous rock is ringed by younger, tilted sedimentary strata. Smaller domes in the Black Hills are laccoliths.

A central basin may develop in a dome as erosion cuts off the top, exposing weaker rock which then erodes relatively fast. Usually streams run down the sides toward the center, joining mainstreams which follow valleys cut through the basin's sides. On the dome's outer slopes some streams run toward the edges; others follow homoclinal valleys in weak belts ringing the dome. Examples: Tennessee's Nashville Dome and the Weald in southern England.

Salt domes are produced by injection of salt in a plastic state into surrounding and overlying beds. Topographic expression results if the salt reaches the surface. In arid regions like Iran the salt mass itself is broadly dome-shaped. In humid lands, removal of the top by solution may leave a shallow depression.

Erosion of structural dome has left stumps of folds rising as hogbacks around exposed sedimentary core. (Rocky Mts, Wyo.)

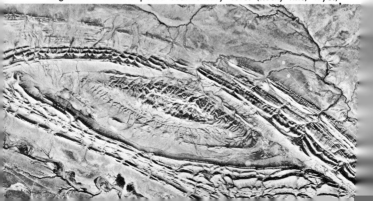

Red Canyon fault scarp was produced at time of Hebgen Lake earthquake in August 1959. (W. Yellowstone, Mont.)

Normal fault is revealed by off-set strata of different hues. Footwall at left, hanging wall at right. (Sandia Mts, N. Mex.)

FAULTING AND FAULTS

Faults are fractures involving displacement of adjoining crustal masses. When a faulting movement occurs, it is generally accompanied by an earthquake. Ordinarily the displacement is only inches at one time; the greatest known was 40 ft., in the Alaskan earthquake of 1964. However, a series of slight displacements over millions of years can produce great mountain ranges and vast depressions. The largest faults are hundreds of miles long and reach down to the mantle.

Fault patterns can be looked for in bedrock exposures. In roadcuts, quarries, and cliffs in layered rock, abrupt offsets along a line indicate a fault. Where such evidence is lacking, a fault may be recognized by mechanical features resulting from the movement, e.g. rock broken or pulverized along a line, or a plane surface polished or gouged. In areas where faulting has occurred relatively recently, displacements may be apparent in the relief—e.g. as angular, steep-sided highlands and depressions. In the United States, such areas are found generally westward from the Rockies.

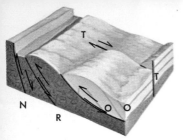

Fault types classified as to relative motion: N normal (gravity); R reverse (thrust); O overthrust; T tear (strike-slip).

Slickensides: polished grooves due to abrasion on fault surface (Hudson Highlands, N.Y.)

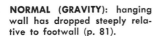

FAULT TYPES

NORMAL (GRAVITY): hanging wall has dropped steeply relative to footwall (p. 81).

REVERSE (THRUST): footwall has dropped steeply relative to hanging wall. Hanging wall has ridden up on footwall.

OVERTHRUST Displacement over many miles at low angle (p. 85).

TEAR (STRIKE-SLIP): movement mostly horizontal along steeply dipping fault plane. Stream or fold crossing fault line may be displaced (p. 83, top right).

RELIEF PRODUCED DIRECTLY BY FAULTING The exposed side of an up-thrown block is a *fault scarp*. It may be inches or miles high, occurring above or below sea level. A fault that dies out may resume on a parallel line beyond (see diagram, p. 29, top right), creating a ramp joining the up-thrown and down-thrown blocks. Ramps are limited to recent fault scarps; e.g. near Upper Klamath Lake, Oreg. Here the up-throw of a fault block dammed the stream. The lake's outlet is through a low spot in the down-thrown block.

Most scarps high above sea level are old and much eroded. The top is the oldest, most-eroded part; the bottom the youngest, least-eroded part. Accordingly, valleys or gullies in the scarp tend to narrow toward the bottom, like a wineglass. The lower ends may "hang" above the scarp base if there has been sub-

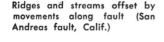
Tilted blocks of Basin and Range type (Tule Mts, Ariz.)

Ridges and streams offset by movements along fault (San Andreas fault, Calif.)

stantial recent upward movement. Fault-scarp remnants between the valleys tend to be triangular (p. 84). Bare rock on the scarp may have polished grooves (slickensides) due to abrasion as movement occurred along the fault plane.

Scarps made by thrusts and overthrusts (diagram, p. 82) overhang, and therefore weather and waste to stable slopes which are opposite to the dip of the fault. Exposed scarps lack slickensides along the fault face.

TILTED BLOCKS: common features, typically with long, gentle backslope and steeper foreslope (fault scarp); e.g. Wasatch Range, Utah (p. 85); San Andres Mts, N. Mex. (p. 94); Teton Range, Wyo. (p. 6). In Europe, Vosges and Black Forest Ranges.

LANDFORMS PRODUCED DIRECTLY BY FAULTING

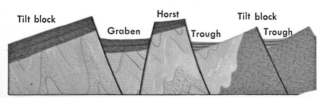

Jordan River graben holds Sea of Galilee. (Tiberias, Israel)

FAULT TROUGHS: elongated depressions due to faulting. One type is along single fault bounding two tilt blocks; trough consists of fault scarp and gentler backslope of a block; numerous in Basin and Range, Nevada. Second type, a *graben*, involves two faults, block being depressed with little tilting; sides are fault scarps. Grabens may be scores of miles long or wide. Some contain lakes, e.g. Lake Baikal in Siberia, world's deepest (5,710 ft.), and lakes in grabens of African Rift Valley. Other grabens, e.g. Rhine depression between Vosges Mts. and Black Forest in Germany, are drained by streams into ocean. Grabens extending far below sea level, with no outlet, include Jordan Valley (Palestine) and Death Valley (Calif.).

Triangular facets are evidence of fault scarp. (Wasatch Mts, Utah)

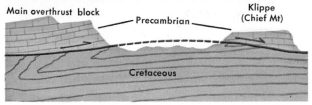

In Lewis overthrust, Precambrian rocks moved east over much younger Cretaceous rocks. Erosion of the overthrust mass finally isolated part of its eastern portion, forming a klippe—Chief Mt.

OVERTHRUST BLOCKS: masses of rock pushed over other masses by low-angle faulting. Movement over scores of miles, as in Alps, tends to bring older rocks from below to cover younger ones. Thus, along the Lewis Overthrust (above) in the Rockies, Precambrian rocks lie over Cretaceous; the movement was at least 15 mi. along 200-mi. front. Overthrust blocks are usually highlands, identified by locating the same overthrust fault plane on opposite sides of the highland; tear faults may complete the block outline. Most striking in Appalachians is westward-moved Cumberland Mountain block (25 mi. wide, 125 mi. long) in Virginia, Kentucky, and Tennessee. Here Pine Mountain Overthrust, trending northeast, is sharply truncated by two tear faults trending northwest. Difficult to see, but shows up well on "topo" maps and from air.

Mythen Peaks (on skyline), at northern edge of Swiss Alps, are klippes far from place of origin. (Lake Luzerne, Switzerland)

KLIPPES: parts of overthrust blocks, isolated by erosion from main mass. They are difficult to distinguish from outliers (isolated erosional remnants) on plains and plateaus. Chief Mt, northwestern Montana, is part of Lewis Overthrust. Smaller klippes are associated with an overthrust westward from Hudson Highlands, N.Y. Matterhorn in Europe is a klippe.

HORSTS: uplifted blocks that are bounded on two sides by fault scarps. Relatively rare.

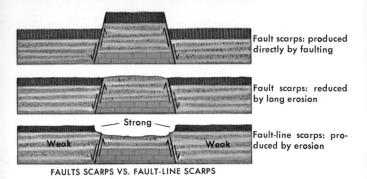

Fault scarps: produced directly by faulting

Fault scarps: reduced by long erosion

Strong

Weak   Weak

Fault-line scarps: produced by erosion

FAULTS SCARPS VS. FAULT-LINE SCARPS

RELIEF INFLUENCED, BUT NOT MADE, BY FAULTS

After relief produced by faulting has eroded away, streams continue to exploit fault lines as belts of breakage and weakness. As stumps of fault blocks become exposed to differential erosion, fault-produced relief is converted into erosional relief. This characterizes landscapes on relatively old fault blocks. Here many hills and valleys correspond to fault lines. Some slopes are *fault-line scarps;* i.e. scarps that follow fault lines but, because relief is erosional, may not face the same way as the original *fault scarp.* A fault-line scarp is less likely than fault scarps to have wineglass valleys and triangular facets, because the scarp emerges in pace with erosion, not relatively rapidly by displacements. The older Appalachians—Blue Ridge, New England Upland, etc. (p. 13)—have many such scarps.

Fault-line scarp made by erosion is on ancient granitic rock. Lowland (left) is on sandstone, which is younger. (Ramapo Mts, N.Y.)

MOUNTAINS

Mountains result from intense deformation of crustal belts. Some mountains are structures formed by folding and faulting; others were built up by series of volcanic eruptions related to the crustal deformation. At the time we see mountains they may be very high, or low but on the way up, or reduced by long erosion.

OROGENY, the development of mountains, starts with accumulation of sediments in a large, sea-filled down-warp known as a *geosyncline*. This may be 1,000 mi. long, 300 mi. or so wide. Over millions of years its bottom sinks as sediments accumulate, perhaps 7 or 8 mi. deep. Then the depressed crust and sediments are melted at depth; metamorphism and igneous activity occur in the geosyncline and along its edges. Meanwhile overlying sediments are folded and thrust-faulted by compression; the entire mass rises; tensions due to uplift result in normal faults. Thus mountains are born.

Orogeny—Phase I: Geosyncline fills and sinks. Igneous activity and metamorphism begin. Phase II: Igneous activity continues. Folding of crust occurs. Uplift makes an island arc (p.61). For Phase III, turn page.

KEY: 1 continent; 2 younger marine sediments; 3 older marine sediments; 4 crust; 5 compression in mantle; 6 melting; 7 ocean basins; 8 zone of later faulting; 9 island arc.

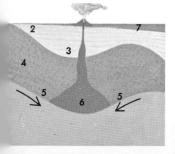
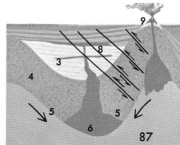

87

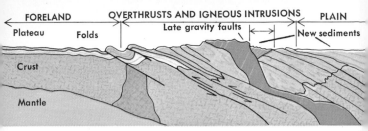

FORELAND OVERTHRUSTS AND IGNEOUS INTRUSIONS PLAIN

Plateau   Folds   Late gravity faults   New sediments

Crust

Mantle

Orogeny—Phase III: Sediments are folded and thrust-faulted by compression. Mass rises to form mountain range.

CRUSTAL ACTIVITY in mountain regions continues over long periods. As mountains are eroded low, others may rise near them. Former mountains may become sites of new geosynclines. The continental cores, or *shields,* consist of igneous and metamorphic rocks that may be roots of the earliest mountains.

One reason for uplifts is erosion. Erosion of a mountain block's top makes the block rise (p. 11). This tendency wanes as the mountain root nears the level of surrounding blocks of about the same density.

Steeply tilted limestone strata, the truncated limbs of great folds, form these cliffs in the Alps. (Grande-Chartreuse, France)

Typical long, nearly level ridges in Folded Appalachians are visible beyond upper part of V-valley in foreground. (W. Virginia)

MOUNTAIN TYPES

Mountains are of four broad classes: foreland-fold, complex, fault-block, and volcanic mountains.

FORELAND-FOLD MOUNTAINS consist of regular folds with occasional thrust and tear faults (p. 82). Faults increase in number and displacement toward the geosynclinal center.

The Folded Appalachians consist mostly of long, parallel, zigzag ridges developed on folds, which change plunge (i.e. undulate) lengthwise. Patterns of synclinal ridges and breached anticlines (p. 79), as seen from the air, resemble canoes in echelon. Their present greatest elevations, about 3,500 ft., with a relief of some 2,000 ft., results from long erosion of alternating beds of differing resistance. Thin sandstone and conglomerate beds alternate with thick, weaker limestone and shale beds. Original folding ended with the Paleozoic (p. 18); several uplifts followed.

Backs of anticlines form many summits of Middle Rockies. Flattish tops may be pediment remnants (p. 124). Longs Peak, Colo., left.

The Middle Rockies are the Rockies of Colorado and Wyoming. The typical ranges are broad anticlines separated by synclinal basins, in a belt trending northwesterly along the axis of the Rocky Mountain geosyncline. Folding and some thrusting produced mountains from the geosyncline at the close of the Mesozoic (p. 18). Long erosion of the anticlines has exposed cores of older granites and metamorphics. Rock waste now occupies the synclinal valleys and forms the adjacent High Plains. Late Cenozoic uplift started or re-

Stumps of tilted sedimentary strata, shaped like flatirons, flank granite ridges in Middle Rockies. (Near Glenwood Springs, Colo.)

Anticlinal ridges and synclinal valleys form mountains of Kerry near Killarney, Eire. These resemble Folded Appalachians.

newed cutting of the present valleys, including many water gaps. Paleozoic and Mesozoic beds form hogbacks or so-called "flatirons" flanking granitic cores, e.g. in foothills of the Colorado Front Ranges. Here the higher summits, including Pikes Peak (14,110 ft.), are granite.

The Jura and Chartreuse region of southern France consists of foreland-fold mountains of the Alpine orogeny. They ceased forming in the late Tertiary (p. 18). Their rather regular folds, occasionally thrust and torn like Appalachian folds, are younger and higher (to 6,000 ft.), with more relief (to 4,000). Limestone masses form most of the high peaks (p. 88).

The Mountains of Kerry in Ireland, like the Appalachians, consist of folded Paleozoic strata. Ridges are stripped anticlines on resistant Devonian sandstones (Carrantuo Hills, 3,414 ft.). Valleys are synclines floored with Carboniferous limestone. Fold trends are east-west. Partial drowning of valleys has made a shoreline with many bays.

Intense folding of sedimentary rock (as seen in distant slope) characterizes many parts of Front Range of Rockies in the United States and Canada. (Glacier Nat. Pk., Mont.)

COMPLEX MOUNTAINS develop near a geosyncline's center (p. 88). Here folds and faults become more complex; recumbent folds and large overthrust slices are typical; rocks become metamorphosed and frequently injected with magma. Examples: Andes and Alpine-Himalayan system—both of recent origin.

Commonly, complex mountains are bordered by foreland folds or plateaus (pp. 104-105), from which sediments are stripped to expose a "basement"—often a worn, older complex mountain mass. The Hudson Highlands and Blue Ridge are complexes of worn-down Precambrian rocks buried by sediments that were deformed in the Appalachian orogeny and later partly removed. Complex mountains may include isolated recumbent folds, exposures of overthrusts, and large igneous masses. On these various structures characteristic relief forms may develop (see facing page).

Laurentian Hills of Canada are worn, complex granitic cores of a mountain range eroded away long ago. (Lac des Sables, Quebec)

BOLD ESCARPMENTS where recumbent folds or overthrusting has brought together rocks of unequal resistance to erosion; e.g. Rocky Mountain Front in Glacier National Park, Mont., (pp. 85 and 92).

LONG RIDGES curving back on themselves, suggesting gently plunging, large folds with original bedding destroyed by metamorphism; e.g. Hudson Highlands, northwest Adirondacks, northern Blue Ridge.

PLATEAU-LIKE MASSES that are underlain by low-dipping beds and may be discovered to be an overthrust on one side of a recumbent fold partly eroded, e.g. Rockies in Glacier Park (p. 85); Ramapo Mts, N.J. p. 86).

CLUSTERS OF IRREGULAR HILLS —usually igneous intrusive masses, e.g. batholiths and stocks. Lack "grain." Some are high by superior resistance, e.g. Mt. Katahdin, Me., and Mt. Blanc, France.

Grampians of Scotland are old, much-worn mountains of complex type, glaciated by Pleistocene ice. (Glen Spean, Inverness)

Fault scarp of San Andres Mts, tilted fault block in Basin and Range Province, rises above Tularosa Basin, a graben. (New Mexico)

FAULT-BLOCK (or BLOCK) MOUNTAINS result usually from gravity faults (p. 82). The process starts when broad uplift stretches the crust. Where the crust's tensional strength is exceeded, fractures occur and the weight of the crustal blocks causes a general collapse with faulting. Differential displacements produce tilt blocks, horsts, and grabens (p. 84).

Block mountains may form in areas where deformation began much earlier. Thus in the Basin and Range area (p. 12) some individual fault blocks show Rocky Mountain folding. Numerous blocks trending north-south lie between super-blocks: the Sierra Nevada on the west and the Wasatch Range on the east. The interior blocks, sunk relative to the super-blocks, make a kind of super-graben. Both escarpments have been eroded but still preserve fault-produced features. The Wasatch fault scarp is 130 mi. long, with summits 4,000 ft. above Great Salt Lake to the west. The Sierra Nevada scarp, 400 mi. long, has summits 10,000 ft. above grabens to the east.

In central England a single tilted fault block runs north-south from the Cheviot Hills to Manchester. Its scarp, forming the western slopes of the Pennine Hills, is the most prominent topographic feature in England, especially at Crossfell Edge, where the Vale of Eden parallels it.

On the European continent the Vosges Mountains and Black Forest regions are tilt blocks associated with Alpine movements. The Vosges scarp faces east toward the opposing Black Forest scarp. The Rhine River follows the intervening graben.

Block mountains may develop in areas of extensive basaltic lava flows, e.g. the Klamath Mountains of California. On ocean bottoms gigantic block mountains parallel the volcanically active rift zones.

VOLCANIC MOUNTAINS build up by long-continued activity around volcanic vents. Such ranges, rising up from sea bottoms, may have been Earth's first lands. Oceanic volcanoes have continued active into the present, e.g. on Iceland, Hawaii, and the Azores. Oceanic volcanoes produce basaltic lavas, mostly from the mantle. Shield forms are typical.

Continental volcanoes produce felsitic and andesitic lavas derived from melting and remelting of conti-

Volcanic mountain grows with new eruption. (Mt. Kiska, Alaska)

nental crust during mountain-building. Such volcanoes occur along Pacific island arcs (Indonesia, Philippines, Japan, Aleutians), in the Atlantic island arc of the Lesser Antilles in the Caribbean, and in Central America—all continental areas undergoing orogeny, as indicated by faulting and earthquakes.

In continents and sea bottoms are rifts from which much basaltic lava erupts. This may be arched into mountains, then surmounted by cones of felsitic composition; e.g. the Cascade Mountains of Oregon and Washington.

Volcanic mountains endure long after becoming extinct. Notable are some of the Cascades (not all extinct), the San Francisco Peaks of Arizona, and cones such as Mt. Taylor in New Mexico. In France a cluster of domes (including Puy de Dome) and cinder cones stand high along a fault line near Clermont-Ferrand. Slieve Gullion and the Mourne Mountains in northeast Ireland are volcanic structures near the famous Antrim lava plateau (p. 108). Africa's most famous extinct or dormant volcano is Kilimanjaro (19,340 ft.), associated with the African Rift Valley.

Glacier Peak, an extinct volcano (background), rises above basement rock of Cascade Mts in western Washington.

Erosion of weak rock layers on Great Plains has created badlands. Note flat surface beyond. (Badlands Nat. Mon., S. Dak.)

PLAINS AND PLATEAUS

Plains and plateaus are terrains that show horizontal layering, even in the deepest valleys. Layers consist mostly of materials eroded from mountains or erupted by volcanoes. In plains the materials have not been shifted far from the deposition sites by uplift or intense deformation, and are poorly consolidated. Uplift that makes plateaus is related to nearby mountain building, which may transform plains into plateaus. Not included among plains and plateaus are areas of greatly deformed metamorphosed and injected (igneous) rocks generally leveled by deep erosion; but these rocks often form plain and plateau basements.

PLAINS

Essential in plains is horizontal layering, without notable crustal disturbance. A plain is close to the level of the last of the deposits that made it. Near the coast this level is sea level, e.g. the Atlantic Coastal Plain. In mid-continent it might be several thousand feet above sea

Floodplain on floor of mature valley. (Chemung River Valley, N.Y.)

level, e.g. Great Plains. Mild warping is likely, perhaps doming, basining, even faulting.

Plains become dissected. Depth of dissection is conditioned by the original altitude of the plain and the time elapsed since deposition formed it. High plains may be little dissected; some may have relief of several hundreds of feet. Coastal plains develop modestly deep valleys near sea level.

ALLUVIAL PLAINS originate by stream deposition. They are often quite smooth, with slopes gently decreasing downstream. Floodplains and deltas are in this group; so are alluvial fans, some of which spread wide and far from the mountains that produced them. Thus in North America the High Plains reach far east from the Rockies, from South Dakota to Texas. In this area fans, floodplains, and lake deposits—erosional materials from the Rockies—started accumulating about 50 million years ago (p. 18). Deposition continued through the Pleistocene. Since then, erosion near the Rockies has isolated the deposits except for a short stretch ("Gangplank") near Cheyenne, Wyo.

Originating in the Rockies and crossing the High Plains are a few widely spaced rivers, e.g. the Platte, Arkansas, Canadian. Broad areas between streams preserve the alluvium's original flatness. Erosional remnants of flat areas stand prominently near principal rivers, e.g. at Scottsbluff, Nebr. Farther north, younger alluvium has been eroded away, and streams are cutting into older lake- or flood-plain sediments, e.g. in the South Dakota badlands (p. 97).

In France a single enormous fan lies on the north flank of the Pyrenees Mountains. It is cut by radial streams that join and flow to the Atlantic.

MARINE PLAINS are shaped from residual marine deposits above water. From Virginia southward, uplifted beaches and a gently curving shoreline evidence previous smoothing by waves and currents. From Miami south to the Florida Keys are reeflike organic deposits which when uplifted present an ungraded surface. Original sea hollows here may be confused with sinkholes (p. 135) formed by solution later.

Marine plain is part of Atlantic Coastal Plain. (Bahia Honda, Fla.)

Snake River

Great Salt Lake

Lake
Bonneville

Utah Lake

Sevier Lake

UTAH

Map shows extent of Pleistocene Lake Bonneville and today's Great Salt Lake. Areas of the old lake bottom now exposed are lake plains.

GLACIAL PLAINS are those shaped largely by continental ice sheets. Ice may smooth terrains of low altitude and relief, and cover them with forms peculiar to glacial work (pp. 118-120).

LAKE PLAINS (lacustrine plains) are sediment-covered bottoms of lakes that went dry because of basin-filling and outlet-cutting, or removal of dams of glacier ice, or climatic change from humid to arid.

Lake plains dot areas glaciated in the Pleistocene (p. 113). Often a north-flowing river's valley was dammed by ice to make a lake, and the lake floor was exposed when the glacier melted. So it was with the 100,000-sq.-mi. Lake Agassiz, which occupied the Red

Plains extending from the western edges of the Wasatch range are parts of Lake Bonneville's bottom. (Logan, Utah)

Lava plain is flat from high fluidity of basalt flows. Older flows are partly covered with vegetation. (Craters of Moon, Idaho)

River Basin from North Dakota and Minnesota north into Canada. Lake plains border the Great Lakes.

In the Great Basin are scores of flat plains that were lake bottoms in the late Pleistocene, when glaciers were melting farther north and the climate was humid. Then a third of Utah was covered by fresh-water Lake Bonneville. Salt carried into the lake by mountain streams left it via its outlet, which went north through Red Rock Pass to the Snake River. With increasing aridity the lake shrank below the level of its outlet; salt no longer left the lake but evaporation continued. Thus the lake became saltier and saltier. Today the vast salt flats and shrunken Great Salt Lake cover lake-plain sediments in the lowest basin areas.

LAVA PLAINS are lands leveled by the filling of valleys by freely flowing lava or freely blowing pyroclastics. From a distance basaltic lava plains, as in northern New Mexico, look very flat. Seen closer, they show details such as spatter cones, pressure ridges, and lava tunnels (pp. 64, 67). Young lava plains near the Snake River, Idaho, e.g. at Craters of the Moon, are little dissected. The Snake River plain is youthful.

Loess usually maintains steep slopes and vertical "columns" where exposed in roadcuts and valley walls. (Eastern Arkansas)

SAND-DUNE PLAINS are made by wind where sand is abundant and vegetation sparse. Such plains are forming now on deserts, beaches, and arid floodplains. On large areas sand may be partly anchored by vegetation (e.g. on Nebraska and Texas coastal plain).

LOESS PLAINS, related to sand-dune plains, are made of wind-blown silt and dust from deserts or semiarid floodplains. This material lies as much as 40 ft. deep on parts of the Midwest. Probably much is from river deposits which, after exposure by glacier melting, were swept by winds before vegetation could cover them. North Eurasia has loess of like origin. Dust for China's loess plains came from the Gobi.

COASTAL PLAINS are sediment-covered areas of continental shelf recently emerged above sea level (p. 142). Typically they are cut by seaward-running, parallel streams in valleys joined at wide angles by tributaries. Some plains end on the landward side in a lowland and cuesta (p. 28) with broad, shallow water gaps. Plains may hold lakes or marshes in shallow basins made by crustal warping or wind erosion. A plain's seaward edge is commonly straight or broadly curving but indented at valley mouths.

ATLANTIC COASTAL PLAIN, New Jersey to Maryland, is on marine sediments interbedded with sediments from Appalachians. It is a *belted,* or *normal,* plain; i.e. a plain with streams that etch out a series of cuestas and lowlands controlled by seaward-dipping layers of varying resistance. From Virginia to Georgia, belts are replaced by terrace patches to form a *terraced* coastal plain. From New Jersey to Cape Hatteras, plain is indented with bays due to warping and to flooding of wide valleys. From Cape Hatteras south, shoreline is smooth but caped (Capes Lookout, Fear, Romain, Kennedy, etc.). At Fear and Romain, large rivers have made deltas, now cuspate (p. 151) from marine action. On Long Island and Cape Cod, coastal plain is masked with glacial deposits.

GULF COASTAL PLAIN is on marine sediments interbedded with erosion sediments from central North America. Eastern and western sections of plain are belted; central section is terraced and split by broad alluvial plain of Mississippi.

COAST OF SOUTHEASTERN ENGLAND has many attributes of belted plain. Cotswold and Chiltern Hills are cuestas. Long inner lowland runs from English Channel to North Sea via Bristol and Nottingham. Oxford and Cambridge occupy lowland between cuestas. Structures related to formation of Alps appear in London Basin and Weald—an eroded-out dome extending across Channel to France. Rather sharp fold ending in Isle of Wight belies term "plain"; area is transitional to plateau.

English Coastal Plain forms a broad lowland bordered by the Cotswold, a cuesta, on skyline. (Near Bath, England)

Appalachian Plateau is cut with deep valleys made by rejuvenated major streams. (New River, Hawk's Nest St. Pk., W. Va.)

SEDIMENTARY PLATEAUS   Uplift of geosyncline deposits (p. 88) of medium thickness forms a sedimentary plateau. Such plateaus have not undergone major deformation and complete removal of sediments, although these may be dissected. The beds were lifted at the same time as the more deformed mountain areas but remain mostly horizontal. Their association with mountain building is shown by broad warps, open folds and monoclines, and faults, and by the presence of marine beds thousands of feet above sea level.

Plateaus include the *foreland* type (p. 88), near the geosynclinal margin (e.g. Appalachian Plateau), and the *median* type, on a rigid block relatively high in the geosynclinal interior (Colorado Plateau). After uplift a low part of a plateau may be covered with sediments that form an overlapping plain, e.g. the West Gulf Coastal Plain on the Appalachian Plateau. Higher parts of the plateau may be eroded to form "typical" plateau topography, with steep out-facing escarpments on the borders, and with mesas and buttes and rock benches in the interior. Drainage is dominantly dendritic, valleys deep, relief locally high.

Very deep erosion of a plateau may reveal the geosynclinal basement, i.e. the original igneous or metamorphic rocks on which the sediments were deposited (p. 88). In the Grand Canyon, basement rock revealed deep in the inner gorge (p. 17) is the stump of old complex mountains. Wider exposures of plateau basements are seen where original sediments were thin and the basement rock very resistant. West of the Adirondack high-peak region, basement rock has been widely exposed by removal of thin edges of the Appalachian Plateau. Another example: the exposed basement of the Massif-Central, south-central France.

Dissection of a sedimentary plateau: typical topography as land is progressively lowered through youth, maturity, and old age.

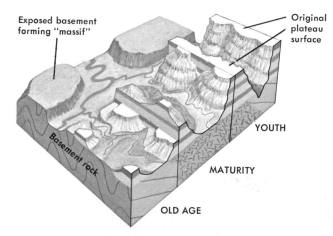

Exposed basement forming "massif"

Original plateau surface

Basement rock

YOUTH

MATURITY

OLD AGE

The Appalachian Plateau (p. 13) was uplifted while the Folded Appalachians were taking form. The plateau strata show mostly broad warping, with strata practically horizontal except in a few localities. In the east the plateau has been carved to form the Catskill "Mountains," in thick, deltaic sediments; the Alleghenies, on very open folds; and the Cumberlands, on an overthrust block. Streams have dissected the plateau, drainage being generally westerly. Major streams of today are the upper Susquehanna and the Allegheny, Monongahela, Kanawha, Cumberland, and Tennessee. Relief, up to 2,000 ft., is often rounded or smooth, expressing humid-climate weathering or glacial erosion. The plateau is bounded mostly by deeply eroded, outfacing escarpments. In the north, glaciation has left magnificent troughs, now occupied by the Finger Lakes, with high waterfalls along hanging valleys. In the south, the Cumberland Plateau has limestone solution features —sinks, lost rivers, etc.

The Colorado Plateau (p. 12) first rose with the Rockies and may have risen again during the past two million years. Incised meanders are common, e.g. the

Rolling terrain characterizes interior areas of Appalachian Plateau that are far from the larger rivers. (Southwestern Pennsylvania)

Rugged canyon scenery has been cut in Colorado Plateau where tributaries descend to meet major rivers. (Colorado Nat. Mon.)

San Juan River near Mexican Hat, Utah, and the Little Colorado near Grand Canyon. Structure is generally horizontal, but with extensive monoclines. Relief often is sharp, in keeping with the dry climate (p. 124). Lava flows often form caprock over sandstone and shale, favoring mesa and butte forms. Near major streams is superb scenery, e.g. at Colorado National Monument, Canyonlands (Utah), Monument Valley and Grand Canyon (Arizona).

LAVA PLATEAUS Volcanic activity over thousands of years may fill a basin with lavas to a depth of thousands of feet. Uplift may make this a lava plateau.

The Columbia Plateau covers about 175,000 sq. mi. of Idaho, Oregon, Washington, Nevada, and California (p. 12). Eruptions since the mid-Cenozoic (p. 18) have deposited lava to depths of 5,000 ft., filling valleys and drowning peaks. (Summits not covered are *steptoes*, e.g. Steptoe Butte, near Garfield, Wash.) Layered lavas are exposed in gorges, as along the

Columbia and Snake Rivers. Some are massive basalt rock, often columned; others are of cinders or colorful (red, yellow, green) tuff or tuff breccia, often welded solid by volcanic heat. Interbedded are sediments deposited by streams between periods of volcanic activity. Well-jointed rock and loose materials contain much ground water, which issues as springs from cliffs. The plateau is laced with dry channels made by melt waters from the Pleistocene ice sheet.

The Antrim Plateau of Ireland, 60,000 sq. mi., is part of a mass of layered lavas that lies mostly beneath the northern Atlantic. The Inner Hebrides, Faroe Islands, Iceland, and East and West Greenland are other above-water parts of this plateau; the remainder has foundered. Fissure flows (p. 67) continue to build up Iceland. Other basalt plateaus: Parana (South America); Deccan (India); areas of East Africa, Australia.

Layers of lava are exposed at a former waterfall and plunge basin bordering Columbia River (foreground). (Dry Falls, Wash.)

Ice cap feeds valley glaciers. (Coast Mts, Alaska-Canada)

GLACIERS AND GLACIATED LANDS

Glaciers make fine scenery and are powerful geologic agents. They whittle mountain peaks, gouge out valleys, grind lowlands. They deplete the ocean, depress the crust, make and fill lake basins, and spread rock waste widely. Today glaciers cover only a tenth of all land— in high mountains and polar regions—but in recent geologic time they have reshaped vast areas.

Glaciers form where annual snowfall exceeds annual ablation (melting and evaporation). Where snow accumulates to about 100 ft. or more, it compacts tightly near the bottom, melts a little and refreezes as temperature fluctuates, takes a grainy form (névé), and gradually turns to solid ice. This, under the overlying weight, spreads out somewhat plastically, like mud. As more snow is added, ice moves out from the center of accumulation as a sheet or stream—a glacier.

VALLEY GLACIERS originate in mountain hollows and move down through valleys previously cut by streams. Movement, usually inches per day, occurs by melting and refreezing within the glacier and by slippage of ice crystals and larger masses against one another. As the ice goes over humps, drags against the valley walls, or spreads where the valley widens, tensions develop, causing vertical fractures called *crevasses*. These, closely spaced, divide the ice into fin- and pillar-like forms called *seracs*. Weathered rock falling onto the glacier from the valley walls forms *moraine*. The ice stream may be joined by *tributary glaciers* from side valleys, each adding another morainal band.

In temperate zones the ice of a valley glacier descends until it reaches a level where ablation destroys it. Here the ice, mantled with rock waste, may terminate as a crumbling wall with an *ice arch,* the opening of a subglacial tunnel from which meltwater pours. This is commonly milky with ground rock ("rock flour," which gives a turquoise hue to many glacial lakes). If the ice front is nearly stationary because of a balance between rate of ice arrival and rate of ablation, a *terminal moraine* (p. 119) accumulates there. In polar lands glaciers enter the sea, producing icebergs.

FEATURES ASSOCIATED WITH VALLEY GLACIER THAT ENTERS SEA

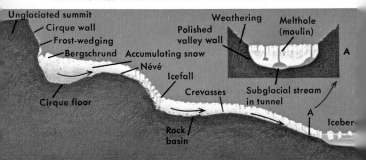

Valley glacier shows contorted morainal bands and intense crevassing. Note lakes on ice margins (lower right), horn peaks, and ice-smoothed rock terrain. (Tikki Glacier, British Columbia)

OTHER TYPES OF GLACIERS

HANGING GLACIERS: ice masses that grow in mountain hollows and terminate at top of steep slope. Ice occasionally spills over edge of hollow and falls or slides to lower levels.

PIEDMONT GLACIERS: broad ice masses formed on lowlands as valley glaciers spread (p. 112).

ICE CAPS: thick ice accumulations that lie as a single mass on a group of highland summits. Valley glaciers may radiate from cap like spokes.

ICE SHEETS: vast ice masses that spread radially from highland centers of accumulation and cover lowlands to depths of thousands of feet. "Tongues" may form on edges. Ice reaches sea through coastal mountains by dividing into *outlet glaciers*. These, because of concentration of flow, discharge into ocean at rates up to 100 linear feet per day. If no mountains bar the way, sheet moves out into water as an *ice shelf*. Ice may be buoyed up and break, forming icebergs.

Expanded-foot glacier, a polar type, is fed by local ice sheet. Glacier front is 30 ft. high. (Commonwealth Glacier, Antarctica)

GLACIERS PRESENT AND PAST

GLACIERS TODAY hold 7 to 8 million cu. mi. of Earth's 325 cu. mi. of water. Ice sheets cover most of Greenland to a maximum 10,000 ft. depth, nearly all Antarctica to 14,000 ft. The weight has depressed central Antarctica a half mile. Ice caps cover mountain areas of Alaska and Norway, and polar islands such as Baffin and Spitzbergen. Piedmont glaciers are restricted to the Greenland, Antarctic, and Alaskan coasts (Columbia Glacier, Prince William Sound). Large valley glaciers are seen in the southern Canadian Rockies, Cascades, Alps and Caucasus, Himalayas, and southern Andes. Small glaciers patch high mountains in temperate regions (e.g. Sierra Nevada, central Rockies, northern Andes) and tropics (Kilimanjaro and other peaks of Africa's equatorial zone).

THE PLEISTOCENE EPOCH (p. 18), the most recent glacial age, began about 1 ½ million years ago as air temperatures dropped about 14°F. below the present average. Ice domes in the north spread to middle latitudes, covering nearly a third of all land. Some lowlands were under ice 2 mi. thick, ice crowned mountains, valley glaciers grew on tropical highlands. Ice-loaded terrains sank as much as 1,000 ft. Sea level was down 200 to 300 ft., exposing continental-shelf areas. As temperatures rose again, melting accelerated. Rock waste from glacial erosion was deposited, ice-made basins filled with meltwater, sea level rose again, and terrains relieved of the ice burden began to "rebound." Then it grew colder, and the process repeated. In all there were four glaciations, each taking 50,000 to 100,000 years. The fourth climaxed about 18,000 years ago; we may now be in an interglacial stage. Possible causes of glacial ages (some occurred earlier in Earth's history) include increased volcanic dust in the atmosphere, reduced solar radiation, and even shifts of the planet's axis of rotation.

Maximum Spread of Pleistocene Ice Sheets

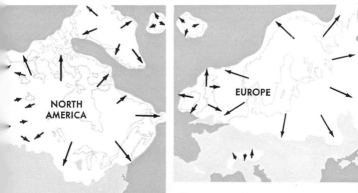

NORTH AMERICA

EUROPE

EROSION BY GLACIERS

Glaciers sculpture land as they move over it. Rock fragments embedded in the ice abrade bedrock and gouge it. The ice breaks off projections. Where a glacier freezes to bedrock, then moves on, it plucks out chunks. Lands exposed by melting of glaciers since the Pleistocene show these effects.

TERRAINS OF LOW OR MODERATE RELIEF: sculptured by ice sheets that covered them. Highlands smoothed and streamlined, e.g. New England, N. Y., N. J.; northern lowlands (Maine, Minnesota) scoured. Shallow basins for ponds plucked out.

MOUNTAINS WITHIN AN ICE SHEET: mostly overridden and smoothed when sheet was thickest. When sheet was thinner in late Pleistocene, valley glaciers enlarged valleys. Examples: Mts. Washington (N.H.) and Katahdin (Me.), each showing both valley and ice-sheet glaciation.

MOUNTAINS OUTSIDE ICE SHEETS: sculptured by valley glaciers. These sharpened peaks, enlarged valleys, dug basins as far south as Sierra Nevada (Calif.) and northern New Mex.

Alpine glacial sculptures here include sharp horn peaks, rock basins, and scoured rock surfaces. (Sierra Nevada, Calif.)

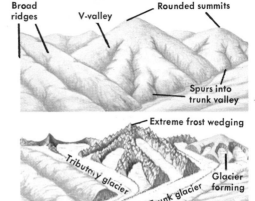

Landscape before glaciation

Broad ridges — V-valley — Rounded summits — Spurs into trunk valley

Landscape during glaciation

Extreme frost wedging — Tributary glacier — Trunk glacier — Glacier forming

Landscape after glaciation (sculpturing more intense toward left)

Horn peak — Col — Arête — Rock basin — Cirques — Truncated spurs — U-valley — Glacial trough — Hanging valley

SCULPTURES BY VALLEY GLACIERS ONLY

The following sculptures are produced only by valley glaciers, because only this type leaves divides exposed.

CIRQUES (English, *corries;* Welsh, *cwms*): hollows at valley heads where valley glaciers originated. Enlarged into amphitheater-like form by plucking and weathering along headwall, and by abrasion of the rock floor. The well-developed cirques have semicircular cliff with basin at foot; basin floor rises toward threshold on downglacier side. After glacier melts basin may hold a pond.

ARÊTES: narrow, jagged ridges between cirques, formed as headwall cliffs erode back.

COLS: low saddles between cirques, formed by intersection of curving cirque walls. Often sites of mountain passes.

HORNS: peaks made by retreat of cirque headwalls from several sides; e.g. Grand Teton (p. 6); Matterhorn in Alps.

U-shaped glacial trough (Near Tracy Arm, Alaska)

U-SHAPED TROUGHS: valleys shaped by moving glaciers into troughs which accommodate flow. When ice melts, trough with U-shaped cross-section (p. 115) is exposed. In valleys not completely ice-filled, upper limit of ice may be shown by a *trim line*, above which valley walls show effects of weathering mainly. All-covering continental ice sheets do not develop trim-line troughs or a fully shaped tributary system.

ROCK BASINS: hollows made by glacial erosion in weaker rock. Many now hold ponds.

HANGING VALLEYS: tributary valleys of relatively low gradient that abruptly terminate in steep wall of large trough. Valley mouths high above trough bottom because downcutting in trough was stronger. Often sites of waterfalls (p. 153).

SCULPTURES BY ICE TONGUES OR BY ICE SHEETS

FIORDS: glacial troughs with floors now below sea level (p. 153). Floors were cut that way or flooded by rising seas during glacier melting.

TRUNCATED SPURS: snubbed-off ends of ridges along sides of valley. Seen in winding valleys somewhat straightened by glacial erosion.

Cirques with small glaciers, horn peaks, truncated spurs, hanging valleys feature Canadian Rockies. (Glacier Nat. Pk., Canada)

STAIRS (ROCK STEPS): found along trough floor, mostly near upper end of main trough at junctions with tributary troughs or at sites of former icefalls.

ROCK KNOBS (French, *roches moutonnées*, "fleece rocks"): resistant bedrock projections; streamlined, often abrasion-polished. Many planed to gentle slope on upstream side, plucked to scarp on downstream side. Hills may be shaped likewise.

Roches moutonnées on granite (Hudson Highlands, N.Y.)

ROCK DRUMLINS: like rock knobs, but with both front and rear ends smoothed (see drumlins, p. 120).

STRIAE (STRIATIONS): grooves or scratches made by rock fragments embedded in ice. Grooves may be more than foot deep, several feet wide, especially in soft, compact rock. Striae give clues to direction of ice movement. Scratches soon weather away if exposed.

Glacial striations preserved on mountain summit (New York)

CRESCENTIC FRACTURES: curved breaks, usually in nested series, made by embedded boulders bumping or "chattering" over bedrock. Up to 12 in. or more long, originally a few inches deep; concave downstream; indicate direction of ice movement.

Crescentic fractures (looking downstream) (New York)

CRESCENTIC GOUGES: depressions made by removal of chips by chattering boulders. Crescent points are on upstream side.

Crescentic gouges (downstream view). (Hudson Highlands, N.Y.)

Terrain Features Near Border of Melting Ice Sheet

GLACIAL DEPOSITS

Much rock waste is produced by glacial erosion. Waste from Pleistocene valley glaciers was left in mountain valleys or washed by meltwater out onto lowlands. Deposits from ice sheets may be found at many levels, even on mountain summits. In the Midwest, rock waste transported by sheets from as far north as central Canada was spread by meltwater, filling valleys several hundred feet deep. Many Northeastern and Midwestern rivers (e.g. the Ohio) were diverted by glacial deposition; some were blocked to make glacial lakes. The Mississippi transported huge masses of outwash to its lower valley and the Gulf. New England valleys are terraced with glacial wastes. Cape Cod and Long Island consist mostly of glacial deposits.

Glacial rock waste is called *drift*—a term used by early geologists who thought it had been dropped by melting icebergs in ancient days when sea covered the land. Drift is sorted and stratified if deposited by water but may be unsorted and unstratified if deposited by ice action. Unsorted drift is *till*.

Terminal moraine holding lake in trough; lateral moraine at right, outwash in foreground (Taylor Valley, Victoria Land, Antarctica)

COMMON DRIFT FEATURES

TERMINAL MORAINES: till accumulations pushed up when glacier front is almost stationary; or drift accumulated in crevasses near front. "Festoons" of drift follow ice front.

LATERAL MORAINES: long, low ridges of mass-wasted rock carried along glacier's sides. Usually scatter as ice melts.

ESKERS: winding ridges of stratified sediments deposited by streams that ran on, within, or beneath a glacier.

KAMES: irregular, rounded, often cone- or dome-like hillocks of stratified drift deposited by meltwater running off glacier sides or into melt-holes. Kames banked along side of trough make *kame terrace*.

KETTLES: depressions in drift due to melting of buried ice blocks. May have ponds.

Ponds occupying scattered kettles in glacial drift (Wisconsin)

"Demoiselles," with protective rock caps: cut by streams in glacial drift (French Alps)

Kame: convenient source for sand and gravel (New Jersey)

Erratic of white conglomerate, carried 40 mi., lies on granite. (Hudson Highlands, N.Y.)

Glacial outwash deposited by meltwater lies in valley below Emmons Glacier, Mt. Rainier, Wash.

DRUMLINS: till molded by ice sheet into streamlined hills, elongated in direction of ice movement. May be 500 ft. high, miles long. Typically steeper in front (upstream side). Often in swarms, suggesting porpoises. Shaped where ice rides over obstacle or extra-thick till (see rock drumlins, p. 117).

TILL PLAINS: deposits completely burying pre-glacial landscape; consist of *ground moraine*, material deposited directly by glacier melting. Clay tills make smooth, undulating plains in Midwest. Sandy and stony till plains less smooth; grade into terminal moraines.

ERRATICS: rock fragments transported by glacier to distant place and deposited as isolated boulders, prominent on broad, smooth surfaces. Often differ from bedrock where found.

VALLEY TRAINS: deposits of sorted drift ("outwash") extending down-valley from moraine in a glacial trough.

OUTWASH PLAINS: sorted drift spread broadly as low alluvial fans fringing an end moraine of a continental ice cap; e.g. Cape Cod, Long Island. Smooth fan may be dotted with kettles, laced with channels of former streams.

Drumlins may occur in swarms. (Central Canada)

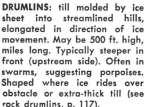

Pillars were sculptured by swift rainwater streams in weak rock of dry, barren terrain. Note resistant caps on pillars. (New Mexico)

DESERT SCENERY

Deserts are terrains with less than 10 in. of rainfall annually: a third of the world's land. Desert climates range from hot, as in North Africa, to cold, as in the Arctic. Lack of rain may be due to (1) remoteness from the ocean, as in Central Asia; or (2) removal of moisture from sea winds by cooling and precipitation as they climb mountain barriers (e.g. Pacific winds that cross the Sierra Nevada before reaching the Great Basin), or (3) heating of air in the planetary circulation as it descends, with increase of pressure, along the horse latitudes (about 30 degrees), remaining dry while passing over land (e.g. African deserts) toward the equator; (4) lack of snow-melting.

Desert land-shaping is dominated by streams despite the general aridity. Rains though rare tend to be sudden and heavy, producing brief torrents. Winds polish and etch rock, and build dunes. Weathering, though limited by the dryness, gradually reduces rock to frag-

Wash leading down from high-lands is filled by a swift, highly erosive stream during heavy rain or melting. Note the steep channel wall. (New Mexico)

ments that become desert pavement, sand dunes, and loess deposits. Rock waste becomes tools for erosion by intermittent streams. Mass wasting goes on, but talus slopes may be minimal because of sudden, effective stream erosion. Mass movements (e.g. earth-flows) that depend on saturation are rare, water tables being low.

During late-Pleistocene meltings (p. 113) lands that are now desert were more humid. Hence older land-forms may show humid influences primarily.

DESERT STREAMWORK

During rain or melting on rocky or hard-packed desert ground, most of the water stays on the surface. Sheets of water sweep the slopes with weathered rock litter. Gathering in gullies, water may become torrents, rushing down toward larger valleys. Torrents become mudflows, or dwindle by evaporation, or are absorbed by seepage on playas (p. 125).

Desert water tables are usually deep; there are few springs to feed streams. Even streams from mountains tend to die out in the desert. A few, such as the Colo-rado and Nile, get enough water from humid mountains to sustain themselves over long desert courses.

DESERT VALLEYS, including the minor trenches called *arroyos,* tend to be steep-sided because the streams are torrential and vigorously abrade the channel with debris. During long periods between rains, alluvium lies exposed on valley bottoms, allowing the wind to winnow sand for dunes and dust for dust storms.

In arid plains, weak rock is dissected by streams to make badland topography. Networks of V-shaped gullies and ravines are typical, as in South Dakota's White River Badlands. Stronger rock favors the canyon with nearly vertical walls, e.g. along edges of the Colorado Plateau in southeastern Utah and in northern Arizona. Here desert streamwork and weathering on rock with pronounced vertical jointing produce pinnacle topography of the Bryce Canyon kind. In some localities meandering streams, rejuvenated (p. 48) by the plateau's uplift, have cut deep, then tunneled through divides between loops to make natural bridges. One of these is Rainbow Bridge, Utah, now high and dry because of continued downcutting.

Rainwash in desert often tends to sculpture high-standing weak material into pointed hillocks. Here clay beds are being eroded down to level of broad resistant layer. (Painted Desert, Ariz.)

Extremely rugged relief is common in desert mountain areas where smoothing effects of weathering are absent. (Big Maria Mts, Calif.)

DESERT HIGHLANDS are commonly sharp in relief, steep-sloped. But as streams cut deeper and grade their beds they tend to swing sidewise, opening out their valleys first near the highland margins and thus helping to form on the bedrock there the broad, gently sloping platforms called *pediments*. These are commonly covered with alluvium (often as fans) from the last period of erosive activity. More recent gullies may expose the eroded bedrock surface. Growth of pediments around a highland may convert it into a feature which is all pediment except for a little residual relief.

Valley wall (foreground) exposes pediment that truncates tilted redbeds. Thin line at top of gullied slopes shows gravel cap on erosion surface (i.e. on pediment). Less-eroded redbeds appear in ridge (middleground). (Wind River Range, Wyo.)

Playas characterize Great Basin lowlands. (Harney Lake, Oreg.)

SOME STREAM-MADE FEATURES

ALLUVIAL CONES AND FANS: well-developed at the foot of many ravines because so much sediment is being transported and ravine usually ends abruptly at a nearly level valley floor. As torrent reaches cone or fan it may divide, cutting separate channels that distribute water and bed load. As flood subsides, channels remain as *washes*, usually dry (p. 122 and below). Water absorbed by cone or fan may emerge as springs at base.

BAJADAS: wide surfaces formed by merging of fans on terrain that lacks outlet streams, e.g. Death Valley, Calif. Bajadas lead gradually down to a *bolson* (Spanish, "purse"), or sump for local drainage.

PLAYAS: low desert plains onto which temporary streams drain, often forming short-lived playa lakes. Very salty ones are *salinas*. Minerals left as lake evaporates make salt or alkali flats.

Fans coalesce to form a bajada. (Mohave Desert, Calif.)

Desert pavement with faceted cobbles (Pacific Creek, Wyo.)

Desert stones with faceting, polish, varnish (Saudi Arabia)

FEATURES MADE BY WIND EROSION

Eddies in steady, gentle winds move sand grains forward on the ground, making *ripple marks*. Stronger, more turbulent winds cause grains to jump one or two feet off the ground—about the upper limit for strong wind abrasion. Undercutting by abrasion is possible along the base of valley walls. Sand by a series of jumps may move up valley walls of gentle slope, e.g. at Hopi Buttes, Ariz. Dust from sand abrasion or weathering may be washed by streams onto playas, then wafted high and far from the source.

PEDESTAL (MUSHROOM) ROCKS: wide-topped pillars cut from rock mass by stream erosion and shaped by rainwash and sandblasting. Narrow base due to intense abrasion near ground. May acquire high polish; layers of differing resistance give grotesque profiles.

VENTIFACTS (Latin, "made by wind"): loose stones faceted by natural sandblasting. Facets commonly triangular. Stones long exposed to prevailing wind may be beveled to point on windward end. Stones that moved may have several points.

DESERT PAVEMENT (SERIR): close-packed, pebble-size ventifacts covering desert bedrock. Too heavy for wind to move.

DESERT VARNISH: high polish on sandblasted rocks, particularly those with coating of iron or manganese dioxide. Common in Death Valley, Calif.

ETCHED ROCKS: rocks with differentially sandblasted surfaces. Softer minerals (e.g. carbonates) are eroded out; harder material (e.g. grains or veins of quartz) remains. Etching may honeycomb rock.

Pillar of gypsum sand held together and stabilized by plant roots in a blowout (White Sands Nat. Mon., N. Mex.)

Natural bridges and arches cut by streams, then smoothed at base by sandblasting (Canyonlands Nat. Pk., Utah)

DESERT WINDOWS: openings cut through narrow, finlike ridges. May resemble natural bridges, but latter usually show evidence of tunneling by stream.

ALCOVES: shallow hollows cut in base of cliffs by sandblasting or stream action. Often enlarged by rockfalls.

BLOWOUTS (DEFLATION HOLLOWS): made by wind work, especially in loose material.

Small mesas, pedestal rocks, and pillars of sand or clay held together by plant roots may stand in a depression as erosional remnants. Some blowouts, as in North Africa, reach water table and become oases. (Oases occur also where ground water emerges through faults or from between tilted strata.) Blowouts common in California's Mohave Desert and High Plains between Nebraska and Texas. Blowouts familiar also in dune areas along sandy coasts.

Desert grass firmly holds mounds of sand in blowout. (Oregon)

DUNES cover only a small fraction of desert lands. They build up where the terrain is flat, vegetation sparse, sand abundant. Vegetation on a dune may stabilize it, but some dunes migrate almost irresistibly, filling canyons and oases, moving up mesa sides, or burying forests. Dunes form not only on deserts but along sandy coasts and on semiarid floodplains.

High-altitude photo of barchan swarms near Moses Lake, Wash.

Notable North American dune areas include Death Valley and Mohave Desert, Calif.; White Sands, N. Mex.; Great Sand Dunes, Colo.; Atlantic Coastal Plain. Saudi Arabia has 400,000 sq. mi. of "sand sea." Dunes are spectacular in the Sahara, Central Asian, and Australian deserts.

Large barchan was made by prevailing winds from west (left). Human figures give scale. (Great Sand Dunes Nat. Mon., Colo.)

BARCHANS: dunes of crescent form. Grains move up windward side by saltation (jumping), then slide down slip face (lee side), forming slope of 30 to 35 degrees. Slip face is hollowed out by eddies. Air flow at dune sides, less turbulent than in middle, extends sides to make "horns." Barchans usually develop in swarms near edge of sand plain (p. 102) where sand is thinning. Oriented by dominant wind, they migrate with it. Shape may be distorted by obstacles or by changes in wind direction.

PARABOLIC DUNES: somewhat like barchans but with horns longer and pointing to windward. Reshaped from dune complexes as blowouts develop between areas of sand anchored by vegetation.

TRANSVERSE DUNES: long, irregular bands at right angles to prevailing wind; are merged barchans.

SEIFS: long, sharp-ridged dunes in Arabian and Saharan deserts. Parallel to direction of prevailing wind; may extend hundreds of miles; up to 300 ft high in Sahara, 600 ft in Iran. Alleys of bare rock between them are *gassi*.

WHALEBACKS: dunes with broad, rounded backs, seen in Sahara. Parallel to direction of prevailing wind. Flattish top. Largest of all dunes; may be 150 ft. high, 2 mi. wide.

Parabolic dunes on Cape Cod made by prevailing winds from northeast (right) (Truro, Mass.)

Transverse dunes, seen from air, with crests about a half mile to a mile apart (West Pakistan)

Air view: seifs (Saudi Arabia)

"Wishing Well" includes "organ pipes" of flowstone (background) and stalagmites of dripstone (foreground). (Luray Caverns, Va.)

LANDFORMS IN LIMESTONE

Extensive landscapes have been developed on carbonate rocks in areas of the crust that were sea bottoms in the ancient geologic past. Limy and other sediments accumulated on the bottoms as chemical precipitates or organic remains, gradually becoming rock. The solubility of this rock in water containing carbon dioxide accounts for scenery like caverns and sinkholes, dry valleys and "lost" rivers. Such features make up *karst topography* (from Karst, Yugoslavia, where solution features are spectacular). The various carbonate rocks in which karst forms develop are referred to in the following pages, for convenience, under the general term "limestone."

Sinkholes, some holding ponds, in limestone (Nova Scotia)

Solution weathering in limestone (New Braunfels, Tex.)

In most regions of sedimentary rock there are phases of limestone or gypsum exhibiting at least minor karst features. In metamorphic zones marble may show some karst detail. The major karst areas in the United States are southern Indiana, Kentucky, Virginia, and central Florida. Nova Scotia has extensive gypsum areas. Caribbean islands, e.g. Puerto Rico, Jamaica, and Cuba, are capped with coral limestone. England has its famous Ingleborough limestone region in Yorkshire, France its Causse and sub-Alps. Mediterranean coasts from Spain to Greece are rich in karst forms.

Roadcut in limestone shows cross section of sinkhole, joints enlarged by solution, and residual red clay (terra rossa). (Kentucky)

THE SOLUTION PROCESS   Calcium carbonate, principal substance in limestone, is insoluble in pure water but soluble in water charged with carbon dioxide. This gas enters surface and ground water from air and from organic wastes. The charged water dissolves limestone and, to a lesser extent, dolomite. Gypsum is soluble in pure water.

Ground water penetrating rock joints and cavities enlarges them by solution. The closer the joints and more rapid the drainage, the faster the process. The resulting features are best developed under humid climates in well-jointed rock elevated above surrounding terrain. Prominent solution features in deserts are likely to have developed during a humid past (p. 122).

Relatively pure limestones are most soluble. Clayey limestone may be nearly insoluble; so also with chalk, because of lack of joints and bedding planes for circulation. Dolomite is less soluble than limestone generally. Insoluble impurities in rock slow solution.

Thin-bedded limestone with pronounced vertical jointing erodes to "pancake rocks" divided by maze of corridors. (New Zealand)

Lost river enters its tunnel at bottom of a hillside (Indiana)

Opening in limestone hillside is exit of stream tunnel (W. Va.)

"LOST RIVERS"

"LOST RIVERS" (sinking creeks) typically run above-ground, then dive into a sink and tunnel made by solution and proceed underground. Collapse of a tunnel roof forms a so-called *window*, exposing a segment of the stream's underground course. Eventually the stream may emerge at a lower level as a *rise,* or *resurgence.* This may be a *well* along the river's course or a large opening in a valley wall above or at the level of the surface stream into which the lost river empties. Rises are related to the presence of insoluble layers. Familiar in uplands, lost rivers often have poorly graded beds because drainage is multilevel.

A lost river gives rise to a famous well in Silver Springs, Fla. In Marble Canyon, Ariz., emerging underground streams make many waterfalls. In southern Indiana, Lost River in going underground left 22 mi. of meandering channel dry. An underground stream in France gives rise to the fountain of Vaucluse near Avignon, source of the Sorgue River.

Virginia's Lost River dives under Sandy Ridge on the west side and emerges on the east side.

LOST AND FOUND
Here Lost River disappears under Sandy Ridge. Two miles away on the other side of the mountain the stream is "found" again as the head-waters of the Cacapon River. This stream has the Indian name for "Medicine Waters."

Lapiés following bedding planes in tilted limestone strata (Chartreuse, French Alps)

SURFACE SCULPTURES

LAPIES (French), **Clints** (British), or **Karren** (German): expanses of compact limestone bedrock grooved by runoff following joints or bedding planes on steeply dipping strata. Grooves may be very deep, forming rock "fins," and may fill with soil. Uplands and lowlands.

PILLARS (PINNACLES): mostly upland features, favored by vertical jointing. Solution speeds downcutting and may produce flowing contours. Spectacular in Dolomitic Alps, Italy.

SOLUTION PANS: very shallow, wide depressions made by solution. Bottoms may be sealed with clay—insoluble impurities remaining after solution of rock. Uplands and lowlands.

COLLAPSE SINKS: irregular surface depressions made by collapse of cave or tunnel roofs. Upland features, less common than sinkholes.

Chalk has been carved into an array of monoliths. Streamwork here is more effective than solution. (Monument Rocks, Kans.)

Sinkhole was formed by solution and subsidence in gypsum. Pond is approximately 100 ft. deep. (Bottomless Lakes St. Pk., N. Mex.)

SINKHOLES (DOLINES, from Doline, Yugoslavia): deep, funnel-shaped holes of all sizes, made by runoff entering joints, usually at intersections. May contain ponds. Some elevated terrains, e.g. central Florida and Kentucky, are riddled with sinkholes. Southern Indiana has perhaps 300,000.

KARST VALLEYS: common upland valleys with prominent features due primarily to solution. Bottom is often poorly graded because of multi-level drainage; may be dry or may contain segment of a lost river. In England, dry valleys are *bournes. Blind valleys* terminate at downstream end in a blank wall, at foot of which stream goes underground. *Pocket valleys* begin with blank wall at upstream end; stream emerges at foot of wall.

VALLEY SINKS (UVALAS): large depressions formed by merging of sinkholes or collapse sinks.

HAYSTACK HILLS: small conical or pyramidal residual hills in old-age phase of degradation. May be riddled with caves. Well developed in Karst, Yugoslavia (where called *hums*); in the Causse, France; and in coral-capped Caribbean islands—Cuba *(mogotes),* Puerto Rico and Jamaica *(pepino hills).*

Haystack hills on gypsum (Cape Breton I., Nova Scotia)

Grottoes were made by ground water and sea action. (Capri, Italy)

LIMESTONE CAVERNS

Once favored by primitive man for shelter and cere-
mony, limestone caves today strongly attract tourists
and spelunkers. Sizable caves, made always in limestone uplands, result from tens of thousands of years of
solutional activity.

Cave formation usually starts where rainfall percolates through joint intersections to the water table. In
a well-bedded, well-jointed limestone, the maximum
ground-water flow and solution occur along bedding
and joint planes below the water table. With horizontal
bedding, vertical development of a cave is concentrated
at joint intersections, horizontal development along
beds which are locally more soluble. When the water
table falls—e.g. from a decrease in rainfall or from
stream downcutting—upper passages go nearly dry
and new passages are dissolved out below. The nearly
dry levels may then start filling with dripstone (p. 139).

Like many other caves, Carlsbad Caverns, N. Mex., were developed to large size in the Pleistocene, when climate was more humid. Cave growth slowed as the climate became dryer and the water table fell, so that a cavern "room" 700 ft. below the entrance now has a dry floor. Other caves open to tourists, e.g. Mammoth Cave, Ky., are relatively dry in upper levels but have a pond below for the traditional boat ride.

Solution features result from removal of rock:

GALLERIES, GROTTOES, ROOMS: large, often flat-arched openings (p. 136) with level or gently sloping floors. Due to solution in more soluble beds and partial roof collapse.

TUNNELS: smaller, twisting passageways that connect galleries, often at different levels. May have sharply vaulted ceilings with fluting suggesting stream erosion during wet periods (indicated also by miniature floodplains on floor). As erosion lowers land, remnants of tunnel roofs become natural bridges or arches.

WELLS (SHAFTS): large vertical openings connecting rooms at different levels or serving as entrances from surface. May be former sinks or resurgences.

Evolution of limestone landscape: Cycle advances left to right; drainage is right to left. A: Sandstones overlying limestone have limited solution; stream has cut deep valley. B: Sinks lead to caves, which drain to river. C: Older sinks have collapsed into caves. Floodplain on cave floor is from former underground river. D: Solution valley was made by underground stream, which becomes a sinking stream, producing a natural bridge. E: Limestone has been removed nearly to level of insoluble layers. A few sinks and sinking streams remain; also haystack hills.

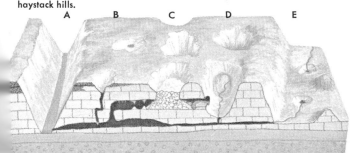

A   B   C   D   E

"Soda Dome" of travertine from warm springs is breached by river. Similar travertine masses accumulate in caves. (New Mexico)

Deposit features in caves take many forms. As water emerges from openings in cave walls and ceilings, some of the carbon dioxide that has held mineral in solution escapes, and carbonate is deposited. (Little deposition occurs from evaporation; caves are mostly too damp.) Deposits, varied and fantastic, are sheltered from outdoor conditions that would destroy them.

TRAVERTINE: familiar rock of limestone caverns; essentially calcium carbonate. White and somewhat translucent when pure; impurities tint it with red, brown, or orange. Artificial lighting adds to color effects.

FLOWSTONE: travertine deposited by flowing water. Flow from horizontal cracks in walls makes "draperies" or "stonefalls." Slow streams from row of holes in wall may account for "organ pipes"; flow over a floor may make terraces. "Lily pads," or "scallops," are made by sheet flow over broad surfaces. Stone may be banded from interruptions in growth or changes in impurities.

Flowstone suggesting a row of grotesque hands (Timpanogos Cave Nat. Mon., Utah)

"Scalloped" surface due to sheet flow over broad surface. (Lehman Caves Nat. Mon., Nev.)

DRIPSTONE: travertine deposited by drip. "Icicle" forms hanging from ceilings are *stalactites;* cones or pillars formed on floor by drip from stalactites are *stalagmites.* Meeting of stalactite and stalagmite makes a *column.* Stalactite growth begins with deposit of ring of mineral on ceiling. This lengthens to a tube, through which water seeps. Eventually the tube is blocked by deposition; thenceforth water creeps down sides. If dripping is relatively rapid, sizable stalactites and stalagmites can form in decades. Largest specimens are tens of thousands of years in making.

Tip of stalactite showing cross section of tube that filled as stalactite grew larger (Kentucky)

Stalactite draperies converging downward to join stalagmites (Lehman Caves Nat. Mon., Nev.)

Young stalactites bristling with helictites (Timpanogos Cave Nat. Mon., Utah)

"Sprays" of aragonite crystals (Timpanogos Cave Nat. Mon.)

HELICTITES: deposits that begin like stalactites, then contort grotesquely. Water may rise from tip by capillary action, somewhat deflected by air currents.

CALCITE or ARAGONITE CRYSTALS: from solutions emerging from pores in walls and ceilings. May be in spongelike masses or clusters of needles. Under cave lights suggest stars.

Gypsum rosette adorning cave floor (Mammoth Cave, Ky.)

GYPSUM ROSETTES: groups of crystals, inches wide, on relatively dry walls and ceilings. Formed as water containing calcium sulfate evaporates within rock pores near surface. Crystal growth pushes crystals to surface, forming "blisters" which break to make the rosettes.

BOXWORK: poorly soluble mineral (e.g. crystalline calcite) that was deposited in joints or minor fractures in limestone and remained as a set of regular, interconnected "boxes" after removal of surrounding limestone by solution.

Gypsum crystals twining like plants (Mammoth Cave, Ky.)

Boxwork on a cave ceiling (Wind Cave Nat. Pk., S. Dak.)

On a shore of submergence, sea cliffs are being carved into rows of stacks, and rock waste is forming beaches. (Cornwall, England)

COASTS AND SHORES

Where sea meets land, geologic agents work visibly, and landforms may change rapidly. Here we sense the persistence of nature in the vastness of time.

A *coast* is a strip of land bordering the sea, within reach of marine processes and influences. Generally it extends as far inland as tidewater does. Thus in New England the width of the coast is 5 to 50 mi.; its landward border is at the limit of tidewater in the rock-walled estuaries. Around Chesapeake Bay, tidewater penetrates as much as 150 mi. up the estuaries.

A *shore* is the zone within which the water line migrates and sea level fluctuates through tides and storms. When narrow it may be called a *shoreline*. It is the seaward boundary of the coast, the landward boundary of the sea. Coasts are attacked along the shoreline not only by land-based agents that shape landscapes elsewhere but also by waves and currents.

141

Lava flows reaching sea form there a neutral shoreline. (Hawaii)

TYPES OF SHORELINES

NEUTRAL SHORELINES develop on coasts made by some sort of deposition above sea level. Much-indented shorelines are found on birdfoot deltas, since here stream deposition has been more active than wave work; e.g. the Mississippi Delta. Relatively smooth, arcuate (arc-shaped) shorelines characterize alluvial fans built into the sea, or built on coastal areas just before a slight, rapid rise of sea level; e.g. Long Island's south shore. Coasts of glacial deposits develop shorelines with shapes depending on those of the deposits; e.g. the Boston (Mass.) drumlin area. Arcuate shorelines form on fronts of lava flows; e.g. Hawaii.

SHORELINES OF EMERGENCE characterize coasts that consist of sediment-covered areas of low relief recently emerged from the sea. If the sediment was spread evenly by marine agents, the shoreline tends to be straight; e.g. eastern Florida. If deposition was uneven (as by organisms in reef and key areas), the shoreline becomes irregular; e.g. southern Florida.

142

Pacific-type Coast

Atlantic-type Coast

SHORELINES OF SUBMERGENCE develop on well-dissected mountain or massif slopes that became submerged by a relative rise of sea level. The shoreline is along river-valley walls. Where the structural grain is oblique to the shoreline trend, there are estuaries, headlands, and islands. These are *Atlantic-type coasts* —e.g. those of Maine and northwestern Spain. Where grain is parallel to the trend, there are long stretches of straight or gently curving shoreline occasionally broken by estuaries, often with gap openings to the sea. These are *Pacific-type coasts*—e.g. the Oregon coast. Recent uplift of the Maine and Oregon coasts has not been enough to "undrown" the larger valleys; the shorelines are still submergent.

Chalk cliffs on submergent shoreline retreat as storm waves undermine them, cutting bench and truncating dry valleys to make so-called "valleuses." (Sussex coast, England)

WAVES AND THEIR WORK

Waves not generated by wind are responsible for short-time changes of sea level.

OCEAN TIDES: produced by Moon's attraction and propagated by Earth's rotation. One crest about every 12 hours allows wind-generated waves to work over vertical zone.

STORM SURGES: more irregular risings and fallings of sea level in response to large atmospheric pressure systems. Set maximum height at which wind waves are effective.

TSUNAMIS: occasional wave trains generated by sudden earth movements, e.g. submarine faults, slumping, volcanic explosions. Announced by fall of sea level at shore; then wall of water comes in. Extreme withdrawal of water in estuaries, followed by swift surging to high levels, often is very destructive. Tsunamis often are loosely called "tidal waves," though not caused by lunar pull.

OCEAN WAVES are made at sea by wind. As surface water is pushed forward, water rises from below to replace it; thus a nearly circular movement is created. Each water particle moves forward at the surface, down, back, up, and forward again. During each revolution the particle makes a small net gain downwind; but the wave form moves much faster. Wave dimensions are directly proportional to the strength of

Waves break on barrier beach, combining with backwash to move sand leeward. Beach scarp was cut by storm waves (New Jersey).

A breaker smashes against rocks at base of cliff. Disintegrating rock becomes sand for the beach. (Los Angeles, Calif.)

the wind and the distance over which it blows. Wave length and wave speed increase with distance traveled, producing *swell*. Erosion by such waves is limited to a back-and-forth motion in loose sediment where orbits of water particles reach bottom.

BREAKERS are forms taken by ocean waves as they enter shallow water and "feel bottom." Circulation in the wave is interrupted, and the wave "breaks." *Plunging breakers* curl and crash; the most destructive kind, they are produced usually by long swell over shallows against opposing wind. *Spilling breakers*, which crumble gradually, are typically made by steep, wind-backed waves of short length. Breaking waves set up a turbulence which moves water and sediment toward the beach or bench, where the waves dissipate as sheets of water (*swash*). Much of the water returns as *backwash*; some is absorbed in dry sand. Oblique swash and backwash promote *beach drifting*, or movement of sediment along the shore.

145

Uplifted coast shows remnants of former wave-cut bench above present bench on which waves are working. (Otter Crest, Oreg.)

Waves attack a shore by washing, impact, wedging (forcing water into cavities and joints under pressure), suction (due to sudden withdrawal of water), and corrasion (rubbing or knocking of rock fragments against one another). Although limited to a narrow zone, wave action is energetic and continuous—most destructive in storms.

Waves breaking on a sea cliff tend to undermine it, producing overhangs. These break off, and the fragments become corrasive tools for further wave attack. Gradually the cliff retreats, and a rock platform called a *bench* takes form and widens beneath it. Large rock fragments are rounded and reduced to cobbles, pebbles, gravel, sand, and silt, and so a beach (p. 149) forms. Waves breaking keep working the sediment and transporting it, generally downwind.

CURRENTS AND THEIR WORK

Currents pick up and transport sand and other weakly consolidated material, eroding some areas and adding to deposits elsewhere.

TIDAL CURRENTS: As tide rises, water penetrates inland via channels; as tide falls, water withdraws. The greater the fluctuation of water level and the more rapid, the greater the velocity of currents and their erosive effect. Tidal currents shape coastal lowlands beyond reach of breakers.

RIP CURRENTS (Rip Tides; Sea Puss): water from breakers and longshore currents, returning seaward. Currents deep, from surface to bottom; commonly extend hundreds of feet seaward. Waves on them usually weak; water agitated, full of churning sand. Current may carry swimmer far out but exerts no "undertow." (Undertow is rare if it exists at all.)

LONGSHORE (LITTORAL) CURRENTS: run approximately parallel and close to shore. Result mostly from action of waves impinging on shore obliquely. Most erosive of shore currents.

Air photo shows barrier island (right) with waves breaking while on seaward side; lagoon (middle) being filled by tidal currents; mainland with river (left). (Metomkin Island, Va.)

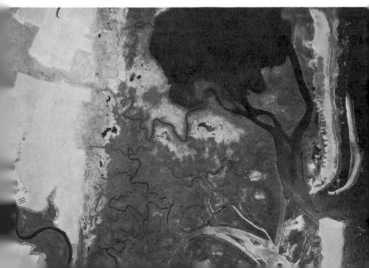

Arches cut through chalk promontories (Normandy, France)

ROCK SCULPTURES

WAVE-CUT NOTCHES: sharp indentations in cliffs made by waves near waterline (photo below). Best developed in headlands by strong undercutting.

SEA CAVES: cavities developed in relatively weak rock in wave-cut notches.

WAVE-CUT BENCHES: rock platforms that extend seaward from base of cliffs (pp. 141, 143, 146, and below). Cut mainly by sediments dragged over bottom by waves and currents, especially during storms. Cutting is at level of storm waves —a little above average level. If coast has recently risen relative to sea level, bench will be above this level; e.g. Pacific Coast, Calif. to Wash. (p. 146).

BLOWHOLES: cave openings through which air compressed by incoming waves blows out. Air may escape through opening in cave roof as wave enters or through cave entrance as water withdraws.

SEA ARCHES: remains of headlands cut through by waves and currents (above, left).

SEA STACKS: columns and pillars left by erosion of wave-cut bench. Favored by vertical jointing. Some are remains of collapsed arches; tops may be remnants of uplifted older benches (p. 146).

SEA GROTTOES: caverns shaped in limestone by ground water and marine action (p. 136).

Sea arch with wave-cut notches stands on bench. (New Brunswick)

Profile of Sand Beach Abutting High Coastal Area

Foreshore: zone between breakers and swash limit. Consists of (1) low-tide bar and (2) trough, made by digging action of breakers; and (3) beach face, made by swash.

Backshore: zone between swash limit and coastal area. Consists of (1) berm, or platform with slight landward slope made by swash as it builds beach seaward; (2) berm crest, or seaward margin of berm built as high swash floods it; (3) winter berm, built by higher winter waves. (Exposed part of winter beach face joins winter and summer berms to make winter-berm terrace.)

BEACHES

Beaches consist of shifting arrangements of sediment along shore. These seaward-sloping deposits are shaped by wave and current action between tidal and storm limits. On rocky coasts, beaches develop along indentations where sediments being moved along shore become trapped.

Beach sediments may come from marine erosion of cliffs or may consist of reworked stream deposits, glacial drift, or volcanic materials. Quartz, feldspar, magnetite, and garnet may be among the minerals present; these with ground-up coral or other shells make up the sand. Various rocks are found as pebbles or cobbles. Flattish cobbles, from thin-bedded rocks, make *shingle*.

The landward border of a beach is either a cliff or a gentle slope beyond reach of all but the highest storm waves.

149

Row of symmetrical beach cusps
(Cape Breton I., Canada)

Swash marks—common features
on sand beaches generally

DEPOSITIONAL SHORE FEATURES

SWASH MARKS: thin, low ridges of sand, seaweed, or other detritus left at highest reach of swash.

RIPPLE MARKS: wavy, approximately parallel ridges of sand or silt molded by waves or currents; exposed at low tide.

RILL MARKS: small drainage channels cut by water from waves or seepage at low tide.

BEACH CUSPS: low mounds, pointed on seaward side, between drainage channels near water level; shaped by waves and currents in tidal zone.

BEACH RIDGES: parapet-like ridges of gravel or cobbles built up by storm waves on landward side of backshore.

SAND BARRIERS: masses of sand carried by waves and longshore currents, deposited usually just above or below tide line where sand is thick, as along coastal plains. Often shift.

BARRIER ISLANDS: broad, long-lasting sand barriers built well above high tide by marine action and by wind. May be breached and reshaped during storms. Called *offshore bars* when far out.

Bay is linked to sea through inlet in baymouth bar. Note small tidal delta. Seaweed darkens water. (Martha's Vineyard, Mass.)

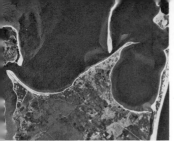

Spits, hooks, and baymouth bars (Martha's Vineyard, Mass.)

Atoll with typical reef-ringed lagoon (Swain's I., SW Pacific)

SPITS: sand barriers connected to shore at one end. Typically form where sand barrier or headland causes formation of longshore currents and deposition of sediments. Spits often curve to a hook or several hooks. Spit grown across mouth of bay is *baymouth bar.*

LAGOONS: areas of relatively quiet water between barrier island and mainland. May gradually fill with sediments washed off barrier by storms, brought in by tidal currents, or deposited by streams off mainland.

TOMBOLOS: sand bars connecting islands to mainland.

CUSPATE FORELANDS: cusp- or wedge-shaped deposits built out from shore at points where obstacle blocks coast-wise movement of sediment, or at eddy points where longshore currents turn seaward; e.g. Capes Kennedy and Hatteras.

ORGANIC REEFS: ridges built up from sea bottom by deposition of hard parts of corals or like organisms. In subtropics. If ridge forms shore, it is a *fringing reef.* Offshore ridge is a *barrier reef,* e.g. Australia's 700-mi.-long Great Barrier Reef. Circular reef rising above water, crowning a submerged volcano, is an atoll.

Tombolo ties sea stack to mainland at left. Jetties were built to trap sand and widen beach at upper right. (Morro Bay, Calif.)

Caves were made by wave action in rock cliffs along shore of a large lake. (Pictured Rocks Nat. Lakeshore, Lake Superior)

LAKESHORES

Lakes generally are short-lived, because they fill with sediment and may empty as their outlet streams cut down. Time is lacking for waves and currents to sculpture the shores. Wave-cut cliffs and terraces may be seen on sediments and relatively weak rock formations along shores of lakes large enough for the growth of sizable waves, e.g. Lake Bonneville (p. 101) and Lake Superior. Shores of kettle lakes (p. 119) tend to be smoothed and rounded by wave and current action. Beaches along large bodies of water, e.g. the Great Lakes, show many features typical of sea beaches.

EVOLUTION OF SHORES

Development of a shore through a characteristic sequence of forms may result from prolonged marine action with little interruption by events such as glaciation or earth movements. The sea tends to straighten the shoreline into broad, sweeping curves. Thus a bird-foot delta would soon become smoothed into an arcuate or cuspate form if the river ceased to flow. But a rocky shoreline tends first to develop cliffed headlands, from which eroded material is swept sidewise and deposited as spits. Continued erosion broadens the wave-cut bench and permits beaches to form. Baymouth bars seal bays, creating a smooth sweep of shoreline at and between headlands.

Shorelines that are initially smooth tend to stay smooth as they retreat. Low-lying emergent shores commonly produce barrier islands in shallow water; these tend to move back toward the mainland and merge with it.

Development of modern shores has been complicated by sea-level changes resulting from glacier melting (pp. 112-113), related isostatic adjustments (p. 113), and non-glacial crustal movements (p. 11). Such events interrupt the wave-current erosion cycle, producing (1) elevated shore features (e.g. California), (2) new initial shorelines (Maine), and (3) submerged shore features (Atlantic shelf).

Fiord coasts show results of Pleistocene glaciation and later rise of sea level. Note trimmed spurs, hanging valley. (New Zealand)

FURTHER INFORMATION

READING TOPOGRAPHIC MAPS

Maps are drawn to scale (1) with minimum shape-distortion and a direction-reference line (2). They show heights and slopes of natural features by contour lines—i.e. closed lines that connect points of the same specified elevation. Some lines close on adjoining maps (3). Points inside a line are higher, those outside lower, than points on the line (4). The opposite is true at closed depressions (5). Nested contours (6) are separated by a stated vertical distance, the *contour interval* (7) although the map distances may vary widely. Non-nested (side-by-side) contours have the same elevation (8). Contours crossing streams indent upstream (9).

Since most landforms are assemblages of slopes, it is important to visualize them. These rules will help:

(a) Slopes descend in the direction of the lowest nested line.
(b) The closer the spacing of lines, the steeper the slopes.
(c) Uniformly spaced nested contours show uniform slopes (10).
(d) Slopes steepen as successive nested contours are successively closer (11) and vice versa (12).
(e) The space between non-nested contours indicates a junction between opposing slopes, i.e. a saddle (8).

Customarily on "topo" maps contour lines are in brown, water in blue, works of man in black, and vegetation in green.

A Topographic Map

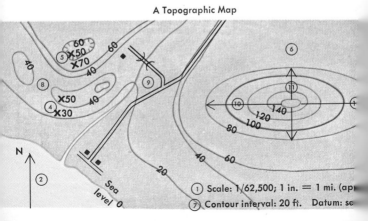

N

① Scale: 1/62,500; 1 in. = 1 mi. (ap
⑦ Contour interval: 20 ft. Datum: se

USEFUL PUBLICATIONS

GENERAL

D. M. Baird, *Guide to Geology for Visitors in Canada's National Parks,* Queen's Printer, Ottawa, Canada.

A. J. Eardley, *General College Geology,* Harper & Row, N.Y., 1965.

A. Holmes, *Principles of Physical Geology,* Ronald Press, N.Y., 1965.

A. K. Lobeck, *Things Maps Don't Tell Us,* Macmillan, N.Y., 1956.

W. H. Matthews III, *Guide to the National Parks* (2 vols.), Natural History Press, Garden City, N.Y., 1968.

John S. Shelton, *Geology Illustrated,* W. H. Freeman and Co., San Francisco, 1966.

W. D. Thornbury, *Principles of Geomorphology,* John Wiley, N.Y., 1969.

J. Wyckoff, *Rock, Time, and Landforms,* Harper & Row, N.Y., 1966.

Zim and Shaffer, *Rocks and Minerals,* Golden Press, N.Y., 1957.

REGIONAL, STATE, AND LOCAL

Information Office, U.S. Geological Survey, Washington, D.C. 20242: Regional geological and "topo" maps, technical bulletins, leaflets for laymen

State tourist information agencies or state chambers of commerce (usually in state capital): State geological maps, road logs, pamphlets.

State geologists (usually in state capital): Technical reports

National Park Service (write to Superintendent of particular Park or Monument): Leaflets and booklets on National Parks and Monuments

American Association of Petroleum Geologists, Box 979, Tulsa, Okla. 74101: Geological highway maps (regional)

Geological Society of America, Box 1719, Boulder, Colo. 80302: Technical studies; field-trip guides

Queen's Printer (or Director, Geological Survey of Canada, Dept. of Energy, Mines, and Resources), Ottawa, Canada: Booklets, leaflets, etc. on Canadian geology.

Various commercial publishers: Regional guide books (in libraries)

INDEX

An asterisk (*) indicates pages on which subjects are illustrated.

158